Annals of Mathematics Study

Number 23

ANNALS OF MATHEMATICS STUDIES

Edited by Emil Artin and Marston Morse

7. Finite Dimensional Vector Spaces, *by* PAUL R. HALMOS
11. Introduction to Nonlinear Mechanics, *by* N. KRYLOFF *and* N. BOGOLIUBOFF
14. Lectures on Differential Equations, *by* SOLOMON LEFSCHETZ
15. Topological Methods in the Theory of Functions of a Complex Variable, *by* MARSTON MORSE
16. Transcendental Numbers, *by* CARL LUDWIG SIEGEL
17. Problème Général de la Stabilité du Mouvement, *by* M. A. LIAPOUNOFF
18. A Unified Theory of Special Functions, *by* C. A. TRUESDELL
19. Fourier Transforms, *by* S. BOCHNER *and* K. CHANDRASEKHARAN
20. Contributions to the Theory of Nonlinear Oscillations, *edited by* S. LEFSCHETZ
21. Functional Operators, Vol. I, *by* JOHN VON NEUMANN
22. Functional Operators, Vol. II, *by* JOHN VON NEUMANN
23. Existence Theorems in Partial Differential Equations, *by* DOROTHY L. BERNSTEIN
24. Contributions to the Theory of Games, *edited by* A. W. TUCKER
25. Contributions to Fourier Analysis, *by* A. ZYGMUND, W. TRANSUE, M. MORSE, A. P. CALDERON, *and* S. BOCHNER
26. A Theory of Cross-Spaces, *by* ROBERT SCHATTEN
27. Isoperimetric Inequalities in Mathematical Physics, *by* G. POLYA *and* G. SZEGO

EXISTENCE THEOREMS IN PARTIAL DIFFERENTIAL EQUATIONS

By Dorothy L. Bernstein

Princeton, New Jersey
Princeton University Press
1950

London: Geoffrey Cumberlege, Oxford University Press
Printed in the United States of America

◇

This study was prepared as part of a survey of large-scale computational problems conducted by Engineering Research Associates, Inc., under contract with the Office of Naval Research, contract N6onr-240.

FOREWORD

This report, "Existence Theorems in Partial Differential Equations" has been prepared for the Office of Naval Research by Engineering Research Associates, Inc., under Contract N6onr-240. Work under this contract included an analysis of information concerning problems involving extensive computations, particularly in connection with military research, in order to determine optimum formulation of problems for solution by computing machines. This list of existence theorems has been compiled to assist those branches of the military which are faced with computational problems involving partial differential equations.

Dr. C. B. Tompkins was in charge of the work under the contract and arranged with Professor Dorothy L. Bernstein of the University of Rochester for the preparation of this report.

H. T. Engstrom
Vice President

27 June 1950

PREFACE

I asked Dr. Bernstein to collect this set of existence theorems during the parlous times just after the war when it was apparent to a large and vociferous set of engineers that the electronic digital calculating machines they were then developing would replace all mathematicians and, indeed, all thinkers except the engineers building more machines.

At that time many Navy activities faced problems of great computational extent, and they naturally examined the power of these machines to solve their problems. The Office of Naval Research had an obligation to advise these activities, and I had an obligation to advise the Office of Naval Research under the terms of contract N6 onr-240 between the Office of Naval Research and Engineering Research Associates, Incorporated. In effect, much of my advice was directly to the activities desiring the equipment.

My task was frequently one of pointing out that the machines will not run themselves. However, I tried to make this advice as constructive as possible, and to do this I frequently tried to estimate the size of machine which, if produced, might be useful in the outstanding problems presented. This, in turn, demanded that we assume a method of solution.

Many of the problems were problems involving partial differential equations. The solution, in many cases, was to be brought about (according to the vociferous engineers) by:

(1) buying a machine;
(2) replacing the differential equation by a similar difference equation with a fine but otherwise arbitrary grid;
(3) closing the eyes, mumbling something about truncation error and round off error; and
(4) pressing the button to start the machine.

One of the simplest problems and one of the best known in which this procedure requires either luck, prayer or further thought is an equation for heat flow:

$$\frac{\partial u}{\partial t} = \frac{\partial^2 u}{\partial x^2}, \qquad 0 \le t, \quad 0 \le x \le 1,$$

with initial and boundary conditions:

$$u(0,x) = f(x),$$
$$u(t,0) = g_0(t),$$
$$u(t,1) = g_1(t).$$

The straightforward difference equation procedure is known to work well on this problem if the grid is properly chosen; however, that arbitrary fineness of the grid does not suffice has long been known. Shortly and verbally, the differential equation defines the following effect: for two adjacent values of t the convexity or the concavity of u, considered as a function of x for these values of t, is decreased during the passage from the earlier t to the later. The rate of flattening depends continuously on the convexity. However, for the obvious difference equation treatment, the flattening for a whole t-step depends on the convexity at the beginning, and unless the t-step is short a convexity may be flattened and the modification continued until a sharp concavity is introduced. The length of the t-step required to do this depends on the square of the length of the x-step, so that an arbitrary refinement of the grid may lead from a good approximate solution to a bad one.

Since these elementary facts seemed to be surprisingly generally unknown, and since neither time nor energy nor budget nor knowledge permitted me to give to each of the Navy's partial differential equations the attention that would be required to apply most of the knowledge already published, this list of existence theorems collected with an eye to computing was obviously desirable to present to the proprietors of problems involving partial differential equations.

In the compilation, Professor Bernstein guided herself by considering the possibility of application to calculation; however, she did not limit herself to theorems that had been proved to be reliable for use on digital computers. This consideration led straightforwardly to the constructive existence theorems contained in the literature which is reasonably easily available. They have been assembled with judgment and analysis; in particular Professor Bernstein believes that the listed theorems have been correctly proved and that they may be applicable or there is some considerable probability that they can be modified so as to be applicable to numerical methods of calculation.

I have been somewhat disrespectful to the optimists. Now, let me warn against extravagant or harmful pessimism. The classical mathematical approach to the study of differential equations is one of proving that a problem has at least one solution, and sometimes of proving uniqueness. There are many times when either an applied mathematician or a pure mathematician may well seek a numerical solution even though he is not certain it is obtainable in the manner chosen (or at all); the purpose may be the application or it may be one of experimental arithmetic - a calculation which it is hoped will help in an effort to guess a correct statement of a proposition which may later be proved. If he believes that there is a considerable positive probability of success and if he can tell whether his final result is an adequate approximation to a solution, there may be no reason for him to demand a firm proof; however, he is obviously foolish if he does not make use of any published material relevant to his problem. The purpose of this collection is to make this relevant material easily available in outline form.

The mission outlined above was not a routine one of copying theorems. The material was drawn from many sources, and not all authors were uniform or careful in defining the problems they attacked. Dr. Bernstein tried to furnish this uniformity by stating what is meant by a solution and by a concise formulation of the various boundary problems encountered. She clarified the hypotheses under which solutions are established, and she carefully defined the regions in which the solutions are valid. She formulated the lemmas connecting various problems and indicated transformations which permit the application of the solution of a given problem to solve related problems. Finally, she included a few definitions which should add greatly to the convenience of most users by eliminating the need for reference to other works.

I believe that these theorems will also have utility to many others. In past years I have spent much time looking for theorems, easily found in this collection, in connection with problems in differential geometry and the calculus of variations. With the idea that this utility might even dominate the utility to computers, I suggested that the collection might be a valuable addition to the Annals of Mathematics Studies.

C. Tompkins

Existence Theorems in Partial Differential Equations

Table of Contents

I Introduction

1. The existence of solutions of partial differential equations is a subject which has engaged the attention of many mathematicians for over a century. During that time the concept of a " solution" and the concept of an " existence theorem" have undergone several modifications. In this chapter, the sense in which these terms are used in this report will be described, and the basis of the classification used for the comprehensive literature which has grown up on the subject will be explained. In addition, certain theorems from the theory of implicit functions and the theory of ordinary differential equations will be stated; they will be needed in later chapters. Geometric terminology is convenient and traditional in discussing certain problems in partial differential equations; since ambuiguity has sometimes resulted when this terminology was not used in the same sense by different authors, certain simple geometric and topological terms will be defined in this chapter, although no attempt has been made to make this an all-inclusive list of such terms.

The n-dimensional Euclidean space E_n is the space of all points P, where P is an n-tuple of real numbers $(x_1, x_2, \ldots, x_n)$. Two points P and $\overline{P}$, with co-ordinates $(x_1, x_2, \ldots, x_n)$ and $(\overline{x}_1, \overline{x}_2, \ldots, \overline{x}_n)$ respectively coincide if, for all i, $x_i = \overline{x}_i$. The distance between P and $\overline{P}$ is the non-negative number $\rho(P,\overline{P}) = \sqrt{\Sigma(x_i - \overline{x}_i)^2}$. For a given $\epsilon > 0$, the ϵ-neighborhood of P is the set of all points $\overline{P}$ such that $|x_i - \overline{x}_i| < \epsilon$ for all i. (One could give a neighborhood definition in terms of $\rho(P,\overline{P}) < \epsilon$; it is essentially equivalent to the one given above.) A set M in n-dimensional space E_n has a limit point L if every ϵ-neighborhood of L contains a point of M different from L. A set M is closed if it contains all its limit points. A set M is open if for every point P in M there is an ϵ-neighborhood of P containing only points of M. The line-segment joining P and $\overline{P}$ is the set of all points Q $(y_1, y_2, \ldots, y_n)$, where $y_i = x_i + (x_i - \overline{x}_i)t$, for $0 \leq t \leq 1$. An open set R such that any two points of it can be joined by a finite number of line-segments lying in R is called a region. Such sets will often be used in what follows; occasionally a "closed region" will be considered; this is a set $\overline{R}$ consisting of a region R and all its limit points.

2. If to every point P of a set M there is assigned a unique real number y, then the set of all such points $(x_1, x_2, \ldots, x_n, y)$ is called a function f with domain M; the value of y corresponding to a particular P is denoted by $f(x_1, x_2, \ldots, x_n)$, or sometimes by $f(P)$. (Thus a real valued function with domain in E_n is a set in E_{n+1}). Let P^o be a limit point of the domain M of f, and let D_δ be a deleted neighborhood of P^o - that is, a set of points P such that $0<|x_i - x_i^o|<\delta$, for all i. Let $D_\delta M$ denote the points common to both D_δ and M. Then $L = \lim_{P\to P^o} f(P)$, or $L = \lim f(x_1, x_2, \ldots, x_n)$, if for every $\epsilon>0$, there exists a $\delta>0$ such that for all points P lying in $D_\delta M$, $|f(P) - L|<\epsilon$. If f is defined at P^o, and $\lim_{P\to P^o} f(P) = f(P^o)$, then f is continuous at P^o. If f is continuous at every point of a set M, then f is continuous in M.

Besides the limits defined above, one has to deal, when $n>1$, with "iterated limits" in certain directions. For example, one could assign fixed values $x_2^o, \ldots, x_n^o$ and consider the function F of the single variable x_1 defined by: $F(x_1) = f(x_1, x_2^o, \ldots, x_n^o)$. If $\lim F(x_1)$ exists, one can write this $\lim_{x_1\to x_1^o} f(x_1, x_2^o, \ldots, x_n^o)$. A similar definition would hold for $\lim_{x_2\to x_2^o} f(x_1^o, x_2, \ldots, x_n^o)$, $\ldots$, $\lim_{x_n\to x_n^o} f(x_1^o, x_2^o, \ldots, x_n)$. From this, one can define successively, $\lim_{x_2\to x_2^o} \lim_{x_1\to x_1^o} f(x_1, x_2, \ldots, x_n)$, ... and finally the "iterated limit" $\lim_{x_n\to x_n^o} \lim_{x_{n-1}\to x_{n-1}^o} \ldots \lim_{x_1\to x_1^o} f(x_1, x_2, \ldots, x_n)$. When $\lim_{P\to P^o} f(P)$ exists, as defined in the previous paragraph, the n! iterated limits all exist and have the same value, but the converse is not true.

When f is a function of a single variable x, $\lim_{h\to 0} \frac{f(x+h)-f(x)}{h}$, when it exists, is called the derivative of f at x and is denoted by $f'(x)$. If the derivative exists for all x in a certain set, this defines a new function which will also called the derivative of f. When f is a function of several variables, the partial derivative of f with respect to x_1 at P will be

$\lim_{h\to 0} \frac{f(x_1+h, x_2, \ldots, x_n) - f(x_1, x_2, \ldots, x_n)}{h}$ when it exists. (Defining F as in the last paragraph, this is just $F'(x_1)$.) It is denoted by $f_{x_1}(x_1, x_2, \ldots, x_n)$, a convenient notation for indicating partial derivatives at particular points. If this partial derivative exists for all x in a certain set, the function thus defined will be denoted by $\frac{\partial f}{\partial x_1}$ or f_{x_1}. A similar definition holds for partial derivatives with respect to any x_i. Higher derivatives are defined in the usual manner, by taking derivatives of derivatives.

$f(x_1, \ldots x_n)$ is said to be of class A^1 *in a region* R if f and all its first partial derivatives are continuous in R. Similarly, f is said to be of class A^m *in* R if f and all its partial derivatives up to and including those of order m are continuous in R. Since the existence and boundedness of all partials of first order in R insure the continuity of a function f, a sufficient condition that f be of class A^m in a closed region R is that all derivatives of order (m+1) exist and be bounded in R. Furthermore, for a function of class A^m, *mixed partials* such as $f_{x_1 x_2}$ and $f_{x_2 x_1}$ are equal; hence, the order of differentiation is immaterial, so we shall use $f_{x_1^{\alpha_1} \ldots x_n^{\alpha_n}}(x_1, \ldots x_n)$ to denote the partial derivative $\frac{\partial^{\alpha_1+\alpha_2+\ldots+\alpha_n} f}{\partial x_1^{\alpha_1} \ldots \partial x_n^{\alpha_n}}$ where $0 \leq \alpha_1+\ldots+\alpha_n \leq m$. Finally, if f can be expanded in an absolutely convergent power series:

$$(2.1) \quad f(x_1, \ldots x_n) = \sum_{\beta_1, \ldots \beta_n = 0}^{\infty} A_{\beta_1 \ldots \beta_n} (x_1 - x_1^0)^{\beta_1} (x_2 - x_2^0)^{\beta_2} \ldots (x_n - x_n^0)^{\beta_n}$$

for all $(x_1, \ldots x_n)$ in N, the region defined by the inequalities $|x_i - x_i^0| < \delta_i$ $(i=1,\ldots,n)$ where $\delta_i > 0$, then we say that f is of class A^∞ in N. It will have continuous derivatives of all orders in N; indeed

$$A_{\beta_1 \ldots \beta_n} = \frac{f_{x_1^{\beta_1}, \ldots, x_n^{\beta_n}}(x_1^0, \ldots, x_n^0)}{\beta_1! \, \beta_2!, \ldots, \beta_n!}$$

3. Given m functions of n variables, f_i, each of class A^1 in a closed region $\overline{R}$, if there exists a function $F(y_1, y_2, \ldots, y_m)$ which is of class A^1 in the entire E_m space, does not vanish in any subregion, and is such that

$$(3.1) \qquad F(f_1(P),\ f_2(P), \ldots, f_m(P)) \equiv 0 \qquad \text{in } \overline{R}$$

then the m functions are said to be functionally dependent in $\overline{R}$. If m functions of n variables are functionally dependent in every closed bounded subregion of a given region R, then the functions are said to be functionally dependent in R. Functions which are not functionally dependent are said to be independent.

Let $f_i(x_1, x_2, \ldots, x_n)$ $(i = 1.2, \ldots, m)$ be of class A^1 in a region R. If $m>n$, these m functions are always functionally dependent. If $m \leq n$, a necessary and sufficient condition that they be dependent in R is that the rank of the " functional matrix"

$$(3.2) \qquad A = \left\| \frac{\partial f_i}{\partial x_j} \right\|_{m \times n}$$

be $r < m$. (That is, r is the largest number k for which there is a determinant of A which is of order k and does not vanish identically in R.) In particular, if m=n, functional dependence can occur if and only if the Jacobian

$$(3.3) \qquad \frac{\partial(f_1, \ldots f_n)}{\partial(x_1, \ldots x_n)} = \left| \frac{\partial f_i}{\partial x_j} \right|$$

vanishes identically in R. (See Kamke[1], pp. 302-311 for proofs.)

Given n functions f_i, each with domain R, to every point $P(x_1, x_2, \ldots, x_n)$ of R there corresponds a point $Q(y_1, y_2, \ldots, y_n)$, where

$$(3.4) \qquad y_i = f_i(x_1, x_2, \ldots, x_n)$$

If the functions are independent in R, then the set of points Q form a region S, such that to every point of S there corresponds exactly one point P of R, given by (3.4). We say that (3.4) defines a one-to-one transformation of the region R onto the region S. The functions $g_i(y_1, y_2, \ldots, y_n)$, with domain S, defined by saying

$$(3.5) \qquad x_i = g_i(y_1, y_2, \ldots, y_n)$$

provided (3.4) holds, are of class A^1 in S, and define a one-to-one transformation of the region S onto the region R, which is called the inverse of the first transformation. The following identities are valid, in R or S:

$$y_i \equiv f_i[g_1(y_1, \ldots, y_n),\ g_2(y_1, \ldots, y_n), \ldots, g_n(y_1, \ldots, y_n)]$$

$$(3.6) \qquad x_i \equiv g_i[f_1(x_1, \ldots, x_n),\ f_2(x_1, \ldots, x_n), \ldots, f_n(x_1, \ldots, x_n)]$$

$$\left[\frac{\partial(f_1, \ldots, f_n)}{\partial(x_1, \ldots, x_n)}\right] \cdot \left[\frac{\partial(g_1, \ldots g_n)}{\partial(y_1, \ldots, y_n)}\right]_{y_1 = f_i(P)} ; \qquad \equiv 1$$

$$\left[\frac{\partial(g_1, \ldots, g_n)}{\partial(y_1, \ldots, y_n)}\right] \cdot \left[\frac{\partial(f_1, \ldots, f_n)}{\partial(x_1, \ldots, x_n)}\right]_{x_i = g_i(A)} \qquad \equiv 1$$

Let f_i be of class A^1 in R, and let (3.4) map R onto a region S; let h_i be functions of class A^1 in S. Then if we set

$$(3.7) \qquad G_i(x_1, \ldots, x_n) = h_i[f_1(x_1, \ldots, x_n), \ldots, f_n(x_1, \ldots x_n)]$$

the functions G_i thus defined are of class A^1 in R and

$$(3.8) \qquad \frac{\partial(G_1, \ldots, G_n)}{\partial(x_1, \ldots, x_n)} \equiv \left[\frac{\partial(f_1, \ldots, f_n)}{\partial(x_1, \ldots, x_n)}\right] \cdot \left[\frac{\partial(h_1, \ldots, h_n)}{\partial(y_1, \ldots, y_n)}\right]_{y_i = f_i(P)}$$

When $n = 1$, these results can be stated as follows: Let $f(x)$ be a function of class A^1 in an open interval R: $a < x < b$, and let $f'(x) \neq 0$ in R. Then the set of points y such that

$$(3.9) \qquad y = f(x) \qquad \text{for } x \text{ in } R$$

form an open interval S: $c < y < d$ (where $c = \lim_{x \to a^+} f(x)$, and $d = \lim_{x \to b^-} f(x)$, if $f'(x) > 0$, but $c = \lim_{x \to b^-} f(x)$ and $d = \lim_{x \to a^+} f(x)$ if $f'(x) < 0$). There is a function $g(y)$ of class A^1 in S defined by setting:

(3.10) $\quad x = g(y)$

whenever (3.9) holds. The function g is called the inverse of the function f. The following identities are valid, in R or S:

(3.11) $\quad x \equiv g(f(x))$; $y \equiv f(g(y))$; $f'(x) \equiv \frac{1}{g'(f(x))}$; $g'(y) \equiv \frac{1}{f'(g(y))}$.

In a great many kinds of problems leading to partial differential equations, one of the independent variables is singled out for special attention; it may be a time-coordinate, as opposed to several space coordinates, or it may be a variable for which an implicit relation is to be solved explicitly. For that reason, we shall take the number of independent variables in a general case as n + 1, denoting the variables by $(x, y_1, \ldots, y_n)$. The simplest case of two independent variables (x,y) will be given special consideration; often a proof is easier to read when we take only two independent variables, and the statement and proof are easily generalized to higher dimensions. Such generalizations are not always possible, as we shall see.

The fundamental implicit theorem referred to above will be applied several times in what follows; it can be stated thus: let $F(x, y_1, \ldots, y_n)$ be of class A^1 in a region S containing the point $(x^0, y_1^0, \ldots, y_n^0)$; let $F(x^0, y_1^0, \ldots, y_n^0) = 0$ and $F_x\ (x^0, y_1^0, \ldots, y_n^0) \neq 0$. Then there is an interval L: $x_1 \leq x \leq x_2$, containing x^0, and a region R containing $(y_1^0, \ldots, y_n^0)$ in its interior such that for every $(y_1, \ldots y_n) \varepsilon R$, the equation

$$F(x, y_1, \ldots, y_n) = 0$$

is satisfied by exactly one x in L. That is, there exists a function $f(y_1, \ldots, y_n)$ with domain R and range L, such that

(3.12) $\quad F(f(y_1, \ldots, y_n),\ y_1, \ldots, y_n) \equiv 0$ in R and such that $x_0 = f(y_1^0, \ldots, y_n^0)$. Furthermore, f is of class A_1 in R, and its derivatives are given by the formulas:

$$f_{y_i}(y_1, \ldots y_n) = -\frac{F_{y_i}(f(y_1, \ldots, y_n), y_1, \ldots, y_n)}{F_x(f(y_1, \ldots, y_n), y_1, \ldots, y_n)}$$

(the denominator does not vanish in $L \times R = S_1 \leq S_2$).

The proof of this theorem is given in standard books of functions of real variables; it is a constructive proof and may be used for the determination of f by computation. When F is given as the limit of a sequence of functions of class A, Germay[1] gave a proof which enables one to determine f by successive approximations. Other theorems on implicit functions relating to the simultaneous solution of several equations can also be found in standard real variable texts. The solution of (3.4) and (3.9), it should be noted, are special cases of implicit function theorems.

4. By a curve C in E_{n+1} is meant a set of points $(x, y_1, y_2, \ldots, y_n)$ such that

$$(4.1) \qquad x = F_1(\mu), \ldots, \; y_n = F_n(\mu) \qquad (\mu \in M)$$

where M is the open interval $a<\mu<b$, and all the functions F_i are continuous in M. If all F_i are of class A^1 in M, and if they do not vanish simultaneously at any point of M, then C is said to be a regular curve.

If $\phi_i(x)$ $(i = 1,2,\ldots,n)$ are functions of class A^1, for x in an open interval M, the set of points $(x, y_1, y_2, \ldots, y_n)$ such that

$$(4.2) \qquad y_1 = \phi_1(x), \; y_2 = \phi_2(x), \ldots, y_n = \phi_n(x) \qquad (x \in M)$$

form a regular curve in E_{n+1}; this is easily seen upon setting

$$(4.3) \qquad F_0(\mu) = \mu \; ; \quad F_i(\mu) = \phi_i(\mu) \qquad (i = 1,2,\ldots,n) \; ,$$

for all the functions F_i are of class A^1 in M, and $F_0'(\mu) = 1$. However the set of points $(y_1, y_2, \ldots, y_n)$ given by (4.2) do not form a regular curve in E_n, unless one adds the condition that $\Sigma[\phi_i'(x)]^2 \neq 0$ for all x in M.

Given a curve C, in its "parametric representation" (4.1), where F_i are all of class A^1 in M, and $F_0'(\mu) \neq 0$ everywhere in M, one can solve the equation $x = F_0(\mu)$ for $\mu = s(x)$ by the methods of the preceding section. Setting

$$(4.4) \qquad \phi_i(x) = F_i(s(x)) \qquad (i = 1, \ldots, n)$$

the points of C are given by (4.2) and all the functions ϕ_i are of class A^1 in M', the x-interval equivalent to the μ-interval M. If any other $F_i'(\mu) \neq 0$ in M, a change of subscripts would lead to a representation of essentially the same character as (4.2). However, if one knows merely that the curve C, given by (4.1) is regular, one can only say that in the neighborhood of each point μ_0 of M, some representation of the form (4.2) or

$$(4.5) \qquad x = \psi_0(y_i), \qquad y_j = \psi_0(y_i) \qquad [j \neq i]$$

is valid, but not that the same representation will hold throughout M.

Thus for plane curves, one may find any of the three representations:

$$x = F(\mu), \ y = G(\mu) \ ;$$

$$y = f(x)$$

$$x = g(y)$$

where all functions involved are of class A^1, The first two are equivalent provided $F'(\mu) \neq 0$; the first and last are equivalent provided $G'(\mu) \neq 0$; the last two are equivalent provided $f'(x) \neq 0$ or $g'(y) \neq 0$.

5. The solution of partial differential equations often involves the solution of ordinary differential equations; a brief summary is given here of basic theorems for systems of first order equations. The presentation follows that of E. Kamke[1] Chapter (IV). Among the many excellent references on the subject of ordinary differential equations, Kamke[1] is recommended as presenting existence theorems in a concise manner which can be readily compared with corresponding theorems on partial differential equations. Chapter II of Kamke[1] gives theorems

for a single first order equation and Chapter VII for a single nth order equation.

Let $f(x, y_1, \ldots, y_n)$ be defined and continuous in a region S. If there exists functions $\phi_1(x), \ldots, \phi_n(x)$ of class A^1 (with continuous derivatives $\phi_i'(x)$), if all $x \in L$, and if

(a) $(x, \phi_1(x), \ldots, \phi_n(x)) \in S$ for $x \in L$

(b) $\phi_i'(x) \equiv f^i(x, \phi_1(x), \ldots, \phi_n(x))$

then we say that

$$(5.1) \qquad y_i = \phi_i(x) \qquad (i=1,\ldots,n)$$

form a *solution*, or an *integral*, of the system of equations:

$$(5.2) \qquad y_i' = f^i(x, y_1, \ldots, y_n).$$

The existence theorems which follow assert the existence of $\phi_i(x)$ which are solutions of (5.2) and are such that: for a given set of numbers $(x_0 y_1^0, \ldots, y_n^0)$

$$(5.3) \qquad \phi_i(x_0) = y_i^0$$

If we consider (5.1) as defining a curve in $(x, y_1, \ldots, y_n)$ space, we call it an integral curve and then (5.3) means that the integral curve is to go through the point $(x_0, y_1^0, \ldots, y_n^0)$.

THEOREM 5.1: Let $f^i(x, y_1, \ldots, y_n)$ be continuous in

$$S_0: \quad |x-x_0|<a, \quad |y_1-y_1^0|<b, \ldots, \quad |y_n-y_n^0|<b$$

where $(x_0, y_1^0, \ldots, y_n^0)$ and a and b are given constants. Let

$$|f^i(x, y_1, \ldots, y_n)| \leq A \quad \text{in } S_0 \qquad (i=1,\ldots,n)$$

$$(5.4) \qquad |f^i(x, \bar{y}_1, \ldots, \bar{y}_n) - f^i(x, y_1, \ldots, y_n)| \leq M \sum_{k=1}^{n} |\bar{y}_k - y_k| \quad \text{in } S_0$$

(Condition 5.4 is called a Lipschitz condition.) Set $\alpha = \min(a, \frac{b}{A})$. Then there exists a unique solution (5.1) of the system (5.2) satisfying (5.3) in

$\bar{L}$: $|x-x_0| < \alpha$. Indeed,

$$(5.5) \qquad \phi_i(x) = \lim_{\lambda \to \infty} \phi_{i\lambda}(x)$$

where

$$(5.6) \begin{cases} \phi_{i_0}(x) = y_i^{\circ} \\ \phi_{i\lambda}(x) = y_i^{\circ} + \int_{x_0}^{x} [f^i(t, \phi_{1,\lambda-1}(t), \ldots, \phi_{n,\lambda-1}(t))]\, dt \quad (\lambda = 1,2,3..) \end{cases}$$

and the convergence is uniform in $\bar{L}$.

Proof: (Picard's method of successive approximation) (Kamke[1], pp. 124-126) Defining the sequences $\{\phi_{i\lambda}(x)\}$ as in (5.6), they are shown to be successively continuous in $\bar{L}$ and such that $|\phi_{i\lambda}(x) - y_i^{\circ}| < b$ when $x \in \bar{L}$ and $\phi_{i\lambda}(x_0) = y_i^{\circ}$, so that $(x, \phi_{1\lambda}(x), \ldots, \phi_{n\lambda}(x)) \in S_0$ and hence we can find f^i at such a point. Then by using the inequality:

$$|\phi_{i\lambda}(x) - \phi_{i\lambda-1}(x)| \le \frac{A}{nM} \frac{[nM\,|x-x_0|\,]^{\lambda}}{\lambda!}$$

and by the comparison test, the convergence of $\lim_{\lambda \to \infty} \phi_{i\lambda}(x) = \sum_{\lambda=0}^{\infty} [\phi_{i\,\lambda} - \phi_{i,\lambda-1}]$ is established, and shown to be uniform in $\bar{L}$. Defining $\phi_i(x)$ as in (5.5), it is shown to possess the desired properties. Uniqueness is established in a later theorem - see p. 141 of Kamke[1].

COROLLARY: The theorem is valid if f^i is of class A^1 in S_0, and indeed if f^i is continuous and each $f^i_{y_k}$ $(i,k = 1,\ldots,n)$ is continuous in S_0.

THEOREM 5.2: Let $f^i(x, y_1, \ldots, y_n)$ be continuous and bounded in:

$$S_1: x_0 \le x \le x_0 + \alpha; \quad -\infty < y_k R +\infty \qquad (k=1,\ldots,n).$$

Then to every set of values $(y_1^{\circ}, \ldots, y_n^{\circ})$ there corresponds an integral curve (5.1) of (5.2) in S_1, passing through $(x_0, y_1^{\circ}, \ldots, y_n^{\circ})$.

Proof: (Kamke[1], pp. 126-128 Due to Peano.)

COROLLARY: Let $F^i(x, y_1, \ldots, y_n)$ be continuous in:

$$S_2: \quad |x-x_0| \leq a, \quad |y_1-y_1^0| \leq b, \ldots, \quad |y_n-y_n^0| \leq b$$

If $|f^i| \leq A$ in S_2, set $\alpha = \min(a, \frac{b}{A})$. Then there exists, in $\bar{L}$: $|x-x_0| \leq \alpha$, an integral curve (5.1) of (5.2) through $(x_0, y_1^0, \ldots, y_n^0)$.

Let $f^i(x, y_1, \ldots, y_n)$ be continuous and bounded in an open region G. Then through every point of G there passes at least one integral curve (5.1) of (5.2), and each $\phi_i(x)$ is of class A^1 in G. Furthermore this curve can be extended up to the boundary of G in both directions: (See Kamke[1], p. 135)

THEOREM 5.3: Let $f^i(x, y_1, \ldots, y_n)$ be of class A^∞ in S_2; define $\alpha = \min(a, \frac{b}{A})$ where, if $f^i : \Sigma A^i_{\alpha,\beta_1,\ldots,\beta_n} (x-x_0)^\alpha \ldots (y_n-y_n^0)^{\beta_n}$, A is defined by:

$$\Sigma \; |A^i_{\alpha,\beta_1,\ldots,\beta_n}| \; a^\alpha \, b^{\beta_1+\cdots+\beta_n} \leq A \qquad (i=1,\ldots,n).$$

Then there exists a unique solution (5.1) of (5.2), such that $\phi_i(x)$ are of class $\overline{A^\infty}$ in $\bar{L}$: $|x-x_0| \leq \alpha$. That is,

$$\phi_i(x) = \sum_{\lambda=0}^{\infty} c_{ij}(x-x_0)^j \qquad (i=1,\ldots,n)$$

where these series converge absolutely for $x \epsilon \bar{L}$.

Proof: Kamke[1], pp. 132. It depends on *method of majorants* which is important, but is omitted here since it will be given in Section 9.

If $f^i(x, y_1, \ldots, y_n)$ and all $f^i_{y_k}$ are continuous in an open region G then through every point of G there passes an integral curve $y_i = \phi_i(x)$ of the system (5.2). If we find the solution corresponding to each point $(\xi, \eta_1, \ldots, \eta_n)$ of G, the solution depends upon the coordinates of this point as parameters - we can therefore write it as:

$$(5.7) \qquad y_i = \Phi^i(x, \xi, \eta_1, \ldots, \eta_n) \qquad (i=1,\ldots,n)$$

That is, going back to (a) and (b) on page 9, for each $(\xi, \eta_1, \ldots, \eta_n) \varepsilon G$, and for each $x \varepsilon L$, (the projection of G on the x-axis),

$$[x, \quad \Phi^1(x,\xi,\ldots,\eta_n), \quad \Phi^2(\ldots), \ldots \quad \Phi^n(x,\ldots,\eta_n) \varepsilon G$$

and

$$(5.8) \qquad \Phi^i_x(x,\xi_1,\eta_1,\ldots,\eta_n) \equiv f^i[x, \Phi^1(x,\xi,\ldots,\eta_n),\ldots,$$
$$\Phi^n(x,\xi,\eta_1,\ldots,\eta_n)].$$

Now considering the functions $\Phi^i(x,\xi,\eta_1,\ldots,\eta_n)$ as functions of $n + 2$ variables defined in the region $G_1 = G \times L$, we call them *characteristic functions* of the system:

$$(5.2) \qquad y' = f^i(x,y_1,\ldots,y_n)$$

and they have the following basic properties:

THEOREM 5.4: Let $f^i(x,y\ ,\ldots,y_n)$ and all $f^i_{y_k}$ be continuous in G. Then the functions $\Phi^i(x,\xi,\eta_1,\ldots,\eta_n)$ defined in (5.5) are of class A^1 in G_1 (i.e., have continuous partial derivatives in all arguments), and at each point of G_1 satisfy the identity:

$$\frac{\partial\Phi^i}{\partial\xi} + \Sigma f^k_i \cdot \frac{\partial\Phi^i}{\partial\eta_k} \equiv 0 \qquad (i=1,\ldots,n)$$

That is,

$$(5.9) \qquad \Phi^i_\xi(x,\xi,\eta_1,\ldots,\eta_n)$$
$$+ \sum_{k=1}^{n} f^k(\xi,\eta_1,\ldots,\eta_n) \cdot \Phi^i_{\eta k}(x,\xi,\eta_1,\ldots,\eta_n) \equiv 0.$$

Furthermore the Jacobian

$$J = \frac{\partial(\Phi^1,\ldots,\Phi^n)}{\partial(\eta_1,\ldots,\eta_n)} \equiv \left|\frac{\partial\Phi^i}{\partial\eta_k}\right| \text{ has the value:}$$

$$J = \exp\left[\int_\xi^x \sum_{k=1}^{n} f^k_{y_k}(t,\phi_1(t,\eta_1,.1.,\eta_k),\ldots,\phi_n(\ldots))\,dt\right]$$

Proof: See Kamke[1], pp. 155-161. The theorem is due to Bendixon and Lindelof.

When the right-hand members of (5.2) do not involve x, the system is:

$$(5.10)\qquad y_i' = f^i(y_1, \ldots y_n) \qquad (i=1,\ldots,n)$$

The characteristic functions then assume a special form; if we set

$$\Phi^i(x,0,\eta_1,\ldots,\eta_n) = \phi^i(x,\eta_1,\ldots,\eta_n)$$

then

$$\Phi^i(x,\xi,\eta_1,\ldots,\eta_n) = \varphi^i(x-\xi,\eta_1,\ldots,\eta_n)$$

Furthermore, if $\sum_\nu f_\nu^2 > 0$, then one can solve the equations

$y_i = \phi^i(x-\xi,\eta_1,\ldots\eta_n)$ to get $\eta_i = \phi^i(\xi-x,y_1,\ldots y_n)$. $(i=1,\ldots n)$

[One often finds differential systems in the literature in the form:

$$(5.11)\qquad \frac{dx}{F_0} = \frac{dy_1}{F_1} = \ldots = \frac{dy_n}{F_n} \qquad (F_i \equiv F_i(x,y_1,\ldots y_n))$$

A special case of this would be:

$$(5.12)\qquad \frac{dx}{F_0} = \frac{dy_1}{F_1} = \ldots = \frac{dy_n}{F_n} \qquad (F_i \equiv F_i(x,y_1,\ldots y_n))$$

which can be considered equivalent to (5.2). That is, given $f^i(x,y_1,\ldots,y_n)$, by a solution of (5.12) one would mean precisely a solution of (5.2). If $F_0,F_1,\ldots,F_n$ are given, and if for some i, $F_i \neq 0$ at every point of S, then by changing the notation, it can be taken as F_0; setting $f^i = \frac{F_i}{F_0}$, (5.11) would be equivalent to (5.12) and hence to (5.2). But if no $F_i = 0$ at every point of S, even though $\Sigma F_i^2 > 0$ in S, then the meaning of a solution of (5.11) in the large becomes less clear, particularly with reference to the properties of characteristic functions. What is usually meant by (5.11) is covered by (5.10) in one higher dimension, and hence we shall use (5.2) and (5.10) rather than (5.11) and (5.12).]

6. Let F(x,y,z,p,q) be a function of 5 variables in a region S: if there exists a function $\phi(x,y)$ of class A^1 in a region R such that for all $(x,y) \in R$,

$$(x, y, \phi(x,y), \phi_x(x,y), \phi_y(x,y)) \in S \text{ and}$$

$$(6.1) \qquad F(x, y, \phi(x,y), \phi_x(x,y), \phi_y(x,y)) \equiv 0$$

we say that $z = \phi(x,y)$ *is a solution of the differential equation*

$$(6.2) \qquad F(x,y,z, \frac{\partial z}{\partial x}, \frac{\partial z}{\partial y}) = 0.$$

[Note the distinction: p is one of the variables for which F is defined; $\phi_x(x,y)$ is a partial derivative of ϕ which may be substituted for p in the expression for F].

We call (6.2) a first order equation in two independent variables and one unknown function.

Similarly, if $F(x, y_1, \ldots, y_n, z, p, q_1, q_2, \ldots, q_n)$ is a function of 2n+3 variables in a region S, and if there exists a function ϕ of class A^1 in a region R of E_{n+1} such that for every $Q:(x, y_1, \ldots, y_n)$ in R,

$$(6.3) \qquad [x, y_1, \ldots, y_n, \phi(Q), \phi_x(Q), \phi_{y_1}(Q), \ldots, \phi_{y_n}(Q)] \in S \quad \text{and}$$

$$F(x, y_1, \ldots, y_n, \phi(Q), \phi_x(Q), \phi_{y_1}(Q), \ldots, \phi_{y_n}(Q)) \equiv 0$$

then we say that $z = (x, y_1, \ldots, y_n)$ is a solution of the differential equation of first order:

$$(6.4) \qquad F(x, y_1, \ldots, y_n, z, \frac{\partial z}{\partial x}, \frac{\partial z}{\partial y_1}, \ldots, \frac{\partial z}{\partial y_n}) = 0$$

One could also consider a system of differential equations of first order. If $F^i(x, y_1, \ldots, y_n, z_1, \ldots, z_r, p_1, \ldots, p_r, q_{11}, \ldots, q_{ik}, \ldots, q_{rn})$, $(i=1\ldots p)$ are p functions of rn+2r+n+1 variables defined in a region S, and if there is a set of functions $\phi^i(x, y\ , \ldots, y\ )$ defined for all $Q:(x, y_1, \ldots, y_n)$ in a region R such that

$$[x, y_1, \ldots, y_n, \phi'(Q), \ldots, \phi^r(Q), \phi'_x(Q), \ldots \phi^r_x(Q), \phi'_{y_1}(Q), \ldots, \phi^i_{y_k}(Q), \ldots] \in S;$$

$$F^i[x, y_1, \ldots, y_n, \phi'(Q), \ldots, \phi^r(Q), \phi'_x(Q), \ldots, \phi^r_x(Q), \phi'_{y_1}(Q), \ldots, \phi^i_{y_k}(Q), \ldots] \equiv 0$$

$$(i=1\ldots p)$$

then we say that $z^i = \phi^i(x, y_1, y_2, \ldots, y_n)$ $(i = 1, \ldots r)$ is a solution of the system of differential equations:

$$(6.5) \qquad F^i(x, y_1, \ldots, y_n, z_1, \ldots, z_r, \frac{\partial z_1}{\partial x}, \ldots \frac{\partial z_r}{\partial x}, \frac{\partial z_1}{\partial y_1}, \ldots, \frac{\partial z_i}{\partial y_k}, \ldots \frac{\partial z_r}{\partial y_p}) = 0.$$

Usually p = r, but this need not be so.

In a similar manner one defines a solution of a single equation of mth order:

$$(6.6) \qquad F(x, y_1, \ldots, y_n, z, \frac{\partial z}{\partial x}, \ldots, \frac{\partial^{\alpha_0 + \ldots + \alpha_n} Z}{\partial x^{\alpha_0} \partial y_1^{\alpha_1} \ldots \partial y_m^{\alpha_n}}, \ldots,) = 0$$

or of a system of partial differential equations of mth order:

$$(6.7) \qquad F^i(x, y, \ldots, y_n, z, \ldots, z_r, \ldots \frac{\partial^{\alpha_0 + \ldots + \alpha_n} Z_i}{\partial x^{\alpha_0} \partial y_1^{\alpha_1} \ldots \partial y_n^{\alpha_n}}, \ldots,) = 0$$

$$\left(\begin{matrix} 0 \leq \alpha_0 + \ldots + \alpha_n \leq m \\ i = 1, \ldots p \end{matrix} \right)$$

(It is important to note that in order to establish that a function ϕ is a solution of a differential equation, one must show that ϕ and its appropriate derivatives exist and *lie in the domain of F*, as well as that they satisfy the equation identically).

An equation such as (6.2) does not have a unique solution - indeed it has an infinite number of solutions. Many papers have been written on the subject of *general solutions*, but there is no simple formulation analogous to that for ordinary differential equations. Another line of investigation has been to determine a function ϕ which in addition to being a solution of (6.2) or (6.4) or (6.6) satisifes certain auxiliary conditions. These auxiliary conditions may be of a great variety of types. (A simple one, for example, in connection with (6.2), is to determine a function ϕ such that $z = \phi(x, y)$ is a solution of (6.2) and that $\phi(x_0,y) = g(y)$, where x_0 is a given point and $g(y)$ a given function of class A^1.) Once these conditions have been stated, the problem becomes that of establishing the existence, and, if possible, the uniqueness, of a function ϕ which is a solution of the differential equation and satisfies the auxiliary conditions. The same thing would apply to systems of equations such as (6.5) or (6.7). It may happen that the auxiliary conditions are such that no solution is possible.) Theorems which prove results of the above type are called existence theorems and form the subject of this paper. A few theorems on general solutions will be given when necessary.

7. Cauchy[1], in 1842, was the first to investigate the existence of solutions of partial differential equations. Since that time, the literature on existence theorems has become very large. Hence a discussion of the results must be accompanied by a classification of the equations which have been studied; such a classification might be done on several different bases:

(I) The number of unknown functions involved, (1,2, or r).

(II) The number of independent variables involved, (1,2,3,4,n.n+1).

(III) The order of the equation or equations, (1,2,3,4,m).

(IV) The nature of the function F (or F^i) which determines the equation; thus for (6.2), one could state concerning $F(x,y,z,p,q)$ that it is of class A^{ω} or that it is linear in p and q, or that it is of the form $p - f(x,y,z,q)$, etc., or that it is defined in a region S of a very particular kind.

(V) The auxiliary conditions involved - both the type of condition and the nature of any functions given in the condition.

(VI) The nature of the desired solution: whether it is to be of class A^1, or A^m, or A^{m+1} or A^∞, what sort of region it is to be defined in, whether it is to be in finite form as an integral equation, an infinite series, etc..

Distinctions made on the first three bases are more obvious and at the same time more trivial than those made on the basis of the last three categories. Often a result may be generalized by increasing the number of independent variables from 2 to rv + 1, or the number of unknown functions from 1 to r, or the order of the equation. Even when this does not happen, a natural question would be to see how far the theorem must be changed in other respects when r or n or m is changed. The division into chapters was made on the basis of III: first order equations are discussed in chapter 2, 2nd order equations in chapter 3, and nth order equations - with some special results for m=3 and m=4 - in chapter 4. A pair of numbers (i,j) after a theorem indicates that it concerns i functions and j independent variables.

Within each chapter, the last three categories form the basis of classification. It becomes apparent that certain kinds of auxiliary conditions (V) are applicable to certain kinds of equations (IV), in the sense that they lead to existence theorems while the same conditions are not applicable to other kinds of equations. Hadamard[1] pointed out this fact in the first decade of this century in connection with the so-called *problem of Cauchy* which has meaning, except for certain special cases, only for *hyperbolic* equations. Hadamard also pointed out the influence of practical problems, in indicating the type of auxiliary condition which would lead to a solution in connection with equations of a given type.

There has been a material change over the last century in regard to (VI), the nature of the desired solution; earlier investigators were concerned only with analytic solutions, or at least those which held in the neighborhood of a given point. Solutions *in the large* are often of very different forms. Whether or not a solution lends itself easily to numerical computation is another criterion which later investigators have considered. Every theorem here will indicate in the hypothesis just what assumptions concerning (IV), (V), (VI) are made - even though they may not have been explicitly stated in the original paper.

II The Initial Value Problem and the Problem of Cauchy for First Order Differential Equations

A. Formulation of the Problem

8. In the space of two independent variables (x,y), and with a single unknown function z, the *Problem of Cauchy* may be stated as follows:

Problem C: (a) Let $x_0(\mu)$, $y_0(\mu)$, $z_0(\mu)$ be functions of class A^1 in $M: \mu_1<\mu<\mu_2$.

(b) Let $F(x,y,z,p,q)$ be a function of class A^0 (continuous) in a region U.

Then it is required to establish the existence of a function $\phi(x,y)$ with the following properties:

(α) $\phi(x,y)$ is of class A^1 in a region R.

(β) For all $(x,y)\ \varepsilon R$, $(x,y,\phi(x,y),\ \phi_x(x,y),\ \phi_y(x,y))\ \varepsilon U$

and

$$(8.1) \qquad F(x,y,\phi(x,y),\ \phi_x(x,y),\ \phi_y(x,y)) \equiv 0.$$

That is, $z = \phi(x,y)$ is a solution in R of:

$$(6.2) \qquad F(x,y,z,\ \frac{\partial z}{\partial x},\frac{\partial z}{\partial y}) = 0$$

(γ) For all $\mu \varepsilon M_1 \leq M$, $(x_0(\mu),y_0(\mu))\ \varepsilon R$

and

$$(8.2) \qquad \phi(x_0(\mu),y_0(\mu)) \equiv z_0(\mu)$$

That is, $z = \phi(x,y)$, considered as a surface, passes through the curve

$$(8.3) \qquad \Gamma: \quad x = x_0(\mu),\ y = y_0(\mu),\ z_0(\mu) \quad (\mu \varepsilon M_1)$$

This problem, stated in such generality, cannot be solved. (See Perron[2]). By adding hypothesis on Γ and F, and by adding requirements to the solution desired, one obtains problems which have been solved. Sometimes two formulations appear quite different, but a solution of one will lead to a solution of the other by a substitution or simple transformation. Several kinds of Cauchy problems will be given, and the next lemma will show when they can be transformed into each other and what the transformations are.

<u>Problem C^m</u>: Same as problem C, except that the functions given in (a) and (b) are of class A^m, for some specified m $(1 \leq m \leq \infty)$.

(A variation of this is to postulate the existence of certain derivatives and then require that they satisfy Lipschitz conditions.)

<u>Problem C_δ</u>: Same as problem C except that M is a δ-neighborhood of some μ, U is a δ-neighborhood of some $(x_0, y_0, z_0, p_0, q_0)$, and the existence of the solution is required only in the δ-neighborhood of (x_0, y_0)

(Such a solution is called a local solution.)

Problem $\bar{C}$: Same as Problem C except that (a) is replaced by $(\bar{a})$ and (γ) by $(\bar{\gamma})$:

$(\bar{a})$ Let g (y) and h (y) be functions of class A^1 in T: $y_1 < y < y_2$.

$(\bar{\gamma})$ For all $y \varepsilon T_1 \leq T$, $(h(y), y)\ \varepsilon R$ and

$$(8.4) \qquad g(y) \equiv \phi(h(y), y) \text{ in } T_1.$$

Problem G: (Initial Value Problem) Same as C except that (a) is replaced by (a_0) and (γ) by (γ_0):

(a_0) Let g(y) be a function of class A^1 in T: $y_1 < y < y_2$ and x_0 be a given number.

(x_0) For all $y \varepsilon T$, $(x_0, y)\ \varepsilon R$ and $g(y) \equiv \phi(x_0, y)$ in T.

Problem I: Same as C except that (a) is omitted and (γ) is replaced by:

(δ) For all $y \varepsilon T$: $y_1 < y < y_2$, $(0, y)\ \varepsilon R$ and $\phi(0, y) \equiv 0$ in T.

Problem N: Same as G except that (b) is replaced by (b_0) and (β) by (β_0):

(b_0) $f(x,y,z,q)$ is of class A^0 in S

(β_0) $z = \phi(x,y)$ is a solution in R of:

$$(8.5) \qquad \frac{\partial z}{\partial x} = f(x,y,z,\frac{\partial z}{\partial y})$$

Problem IN: Same as N, using (δ) in place of (γ_0) and omitting (a).

Problems analogous to $\overline{C}$ - N using g(x) and h(x) instead of g(y) and h(y), and

$$\frac{\partial z}{\partial y} = f(x,y,z,\frac{\partial z}{\partial x})$$

instead of (8.5), do not differ from the above except in notation. Equation (8.5) is called the *normal form* of (6.2).

One can combine these types of problems to get, for example, $\overline{G}^m$, G^m, I^m, N^m or C_δ, G_δ, I_δ, N_δ.

LEMMA 1: Solution of C $\rightarrow$ Solution of $\overline{C}$ $\equiv$ Solution of G

$\equiv$ Solution of I $\rightarrow$ Solution of N $\equiv$ Solution of IN.

Solution of $\overline{C}$ $\rightarrow$ Solution of C, provided $y_0^{(1)}(\mu) \neq 0$ in M.

Solution of N $\rightarrow$ Solution of G, provided F is of class A^1 and $F_p \neq 0$ in U.

Solution of N $\rightarrow$ Solution of $\overline{C}$, provided F is of class A^1 and $F_p - h^{(1)}(y) \cdot F_q \neq 0$ in U.

Solution of N $\rightarrow$ Solution of C, provided F is of class A^1, $F_p \cdot [y_0'(\mu_0(y))] - F_q \cdot [x_0'(\mu_0(y))] \neq 0$ in U and and $y_0'(\mu) \neq 0$ in M where $\mu_0(y)$ is the inverse of $y_0(\mu)$.

(Formulas relating solutions given in proof.)

Proof: Given solution of C, $\bar{C}$ is solved by setting $x_0(\mu)$, $= h(\mu), y_0(\mu) = \mu$, $z_0(\mu) = g(\mu)$, and $M = T$. The solution $z = \phi(x,y)$ satisfies α, β, γ and hence $\bar{\gamma}$, since for $y \varepsilon T_1 = M_1$, let $\mu = y$:

$$\phi(h(y),y) = \phi(x_0(y), y_0(y)) = z_0\ (y) = g(y).$$

Given solution of $\bar{C}$, G is solved by setting $h(y) \equiv x_0$, since this is of class A^∞ in T.

Given solution of G, I is solved by setting $x_0 = 0$ and $g(y) \equiv 0$, since this is of class A^∞ in T.

Given solution of C, N is solved by setting $F(x,y,z,p,q) \equiv p - f(x,y,z,q)$, since if $f(x,y,z,q)$ is of class A^1 in S, $F(x,y,z,p,q)$ is of class A^1 in U, the set in 5-dimensional space having p arbitrary and $(x,y,z,q)\ \varepsilon S$. (We can write $U = L \times S$, L being: $-\infty < p < +\infty$.)

Conversely, suppose solution to $\bar{C}$ is known. Given problem C, with $y_0^{(1)}(\mu) \neq 0$, let $\mu = \mu_0(y)$ be the inverse of $y = y_0(\mu)$ and set $g(y) = z_0(\mu_0(y)$ and $h(y) = x_0(\mu_0(y))$. Then $g(y)$ and $h(y)$ will be of class A^1 (or A^m) if $x_0(\mu)$, $y_0(\mu)$, $z_0(\mu)$ are of class A^1 (or A^m), in an interval T: $y_1 < y < y_2$ corresponding to M: $\mu_1 < \mu < \mu_2$. Solve problem $\bar{C}$, and obtain a function $\phi(x,y)$ satisfying α, β, and $\bar{\gamma}$. For every $\mu \varepsilon M$, there is a unique $y = y_0(\mu) \varepsilon T$ and since $x_0(\mu_0(y)) = x_0(\mu) = h(y)$,

$$\phi(x_0(\mu),\ y_0(\mu)) \equiv \phi[h(y_0(\mu)),\ y_0(\mu)] \equiv g(y_0(\mu)) \equiv g(y) \equiv z_0(\mu)$$

so that γ is verified. Thus $z = \phi(x,y)$ is a solution of $\bar{C}$.

Suppose a solution of I is known, and problem $\bar{C}$ is given. We need to find $\phi(x,y)$ such that (8.1) and (8.4) hold. The transformation:

$$T: \begin{cases} x = \bar{x} + h(\bar{y}) \\ y = \bar{y} \end{cases} \text{ has Jacobian } \begin{vmatrix} 1 & h'(\bar{y}) \\ 0 & 1 \end{vmatrix} = 1$$

and inverse

$$T^{-1}\begin{cases}\bar{x} = x-h(y)\\ \bar{y} = y\end{cases}$$

Under this transformation, each point of R goes into a unique point of $\bar{R}$ and the curve C: $x = h(y)$ $(y\varepsilon T)$ goes into the line $\bar{x} = 0$ $(y\varepsilon T)$.

If we set $\bar{\phi}(\bar{x},\bar{y}) = \phi(\bar{x}+h(\bar{y}),\ \bar{y})$, assuming that $\phi(x,y)$ exists and is of class A^1 in R, then $\bar{\phi}(\bar{x},\bar{y})$ is of class A^1 in $\bar{R}$ and

$$\phi_x(\bar{x}+h(\bar{y}),\bar{y}) \equiv \bar{\phi}_{\bar{x}}(\bar{x},\bar{y});\ \phi_y(\bar{x}+h(\bar{y});\bar{y}) = \bar{\phi}_{\bar{y}}(\bar{x},\bar{y}) \quad -h'(\bar{y}) \quad \cdot\ \bar{\phi}_{\bar{x}}(\bar{x},\bar{y}).$$

Then (8.1) becomes

$$0 \equiv F[\bar{x}+h(\bar{y}),\bar{y},\ \phi(\bar{x}+h(\bar{y}),\bar{y}),\ \phi_x(\bar{x}+h(\bar{y}),\bar{y}),\ \phi_y(\bar{x}+h(\bar{y}),\bar{y})]$$

$$\equiv F[\bar{x}+h(\bar{y}),\bar{y},\ \bar{\phi}(\bar{x},\bar{y}),\ \bar{\phi}_{\bar{x}}(\bar{x},\bar{y}),\ \bar{\phi}_{\bar{y}}(\bar{x},\bar{y}) \quad -h'(\bar{y}) \cdot\bar{\phi}_{\bar{x}}(\bar{x},\bar{y})]$$

Letting $\psi(\bar{x},\bar{y}) = \bar{\phi}(\bar{x},\bar{y}) - g(\bar{y})$, we have $\psi_{\bar{x}} = \bar{\phi}_{\bar{x}}$ and $\psi_{\bar{y}} = \bar{\phi}_{\bar{y}} \quad -g'(\bar{y})$. Thus we get:

$$(8.6)\quad 0 = F[\bar{x}+h(\bar{y}),\bar{y},\ \psi(\bar{x},\bar{y}) + g(\bar{y}),\ \psi_{\bar{x}}(\bar{x},\bar{y}),\ \psi_{\bar{y}}(\bar{x},\bar{y}) + g'(\bar{y}) - h'(\bar{y}) \quad \psi_{\bar{x}}(\bar{x},\bar{y})]$$

Now let

$$\bar{z} = z - g(y),\ \bar{p} = p,\ \bar{q} = q - q'(y)+h'(y)\cdot p$$

and write

$$(8.7)\quad \bar{F}(\bar{x},\bar{y},\bar{z},\bar{p},\bar{q}) \equiv F(\bar{x}+h(\bar{y}),\bar{y},\bar{z} + g(\bar{y}),\ \bar{p},\bar{q} - h'(\bar{y})\cdot\bar{p} + q'(\bar{y}))$$

If F is continuous in a region U, $\bar{F}$ is continuous in $\bar{U}$, the region corresponding to U under:

$$(8.8)\quad \begin{cases} x = \bar{x}+h(\bar{y})\\ y = \bar{y}\\ z = \bar{z}+g(\bar{y})\\ p = \bar{p}\\ q = \bar{q} - h'(\bar{y})\cdot\bar{p} + g'(\bar{y})\end{cases}$$

If we find a solution $\bar{z} = \psi(\bar{x},\bar{y})$ of: $\bar{F}\ (\bar{x},\bar{y},\bar{z}, \frac{\partial \bar{z}}{\partial \bar{x}}, \frac{\partial \bar{z}}{\partial \bar{y}}) = 0$, such that $\psi(0,\bar{y}) \equiv 0$ (Problem I), then $\bar{F}\ (\bar{x},\bar{y},\psi(\bar{x},\bar{y}),\psi_{\bar{x}}(\bar{x},\bar{y}),\psi_{\bar{y}}(\bar{x},\bar{y})) \equiv 0$

But left side is same as right side of (8.6), by definition (8.7), and since (8.6) is equivalent to (8.1),

$$z = \phi(x,y) = \psi(x-h(y),y)+g(y)$$

satisfies α and β. γ is true since $\phi(h(y),y) \equiv \psi(0,y) + g(y) \equiv 0+g(y) \equiv 0$ for y on T. Thus we get G. [Note that if I is solved for functions F of class A^1, and we wish to solve G for functions F of class A^1, then the functions of the transformation (8.8) must be of class A^1 in $(\bar{x},\bar{y},\bar{z},\bar{p},\bar{q})$ which means h(y) must be of class A^2.] Combining this result with the first part of the lemma we get:

$$\bar{C} \equiv G \equiv I.$$

Now suppose N is solved. Given problem G with $F_p \neq 0$ in U, by the implicit function theorem, there is a function of (x,y,z,q) defined in S, the projection of U, such that:

(8.9) $\quad F(x,y,z,f(x,y,z,q),q) \equiv 0$

(If F is of class A^1, so is f; if F is of class A^m, so is f) and

$$f_x = -\frac{F_x}{F_p}, \ldots$$

Solve problem N for this function. There exists $\phi(x,y)$ satisfying α, γ_0 and β_0. That is:

(8.10) $\quad \phi_x(x,y) \equiv f(x,y,\phi(x,y),\phi_y(x,y))$

Substituting $z = \phi(x,y)$ and $q = \phi_y(x,y)$ in (8.9) and using (8.10), we get (8.1). Thus $z = \phi(x,y)$ satisfies β and problem G is solved.

As a consequence of the previous results, it follows that solution of N implies the solution of I, provided $F_p \neq 0$. That $N \equiv IN$ is proved in a manner analogous to that used to show $G \equiv I$.

Suppose N is solved, and problem $\overline{C}$ is given. Define $F(\overline{x},\overline{y},\overline{z},\overline{p},\overline{q})$ as in (8.7). If $\overline{F}_{\overline{p}}(\overline{x},\overline{y},\overline{z},\overline{p},\overline{q}) \neq 0$, we could then find a solution to problem I for $\overline{F} = 0$; i.e., a solution $z = \psi(x,y)$ such that $\psi(0,\overline{y}) \equiv 0$. By (8.7) and (8.8),

$$\overline{F}_{\overline{p}}(\overline{x},\overline{y},\overline{z},\overline{p},\overline{q}) \equiv F_{\overline{p}}(\overline{x}+h(\overline{y}),\overline{y},\overline{z}+g(\overline{y}),\overline{p},\overline{q}\cdot -h'(\overline{y})\cdot\overline{p}+g'(\overline{y}))$$

$$= F_x(x,y,z,p,q)\cdot 0+F_y\cdot 0+F_z\cdot 0+F_p\cdot 1+F_q\cdot(-h'(y))$$

so the condition becomes:

(8.11) $F_p(x,y,z,p,q) - h'(y)\cdot F_q(x,y,z,p,q) \neq 0$ in U.

Solving I as indicated, we need only set $\phi(x,y) = \psi(x-h(y),y)+g(y)$, as in the first part, to get solution to $\overline{C}$.

Again, $N \to C$ if (8.11) holds, and the condition that $[\overline{C}] \to [C]$, namely $y_0'(\mu) \neq 0$. But then since $h'(y) = + \frac{x_0'(\mu)}{y_0'(\mu)}$, (8.11) becomes

$y_0' \cdot F_p - x_0' \cdot F_q \neq 0$. The formulas would be:

Let $\overline{F}(\overline{x},\overline{y},\overline{z},\overline{p},\overline{q}) \equiv F(\overline{x}+x_0(\mu_0(\overline{y}),\overline{y},\overline{z} + z_0(\mu_0(\overline{y})),\overline{p},\overline{q} - \frac{[x_0'(\mu_0(\overline{y}))\cdot\overline{p}+z_0'(\mu_0(\overline{y}))]}{y_0'(\mu_0(\overline{y}))}$

where $\mu_0(y) = \mu$ is the inverse of $y = y_0(\mu)$. Find solution of $\overline{F}(\overline{x},\overline{y},\overline{z},\overline{p},\overline{q}) = 0$ as $p = \overline{f}(\overline{x},\overline{y},\overline{z},\overline{q})$. Get solution of problem IN for $\frac{\partial\overline{z}}{\partial\overline{x}} = f(\overline{x},\overline{y},\overline{z},\frac{\partial\overline{z}}{\partial\overline{y}})$. If this is $\overline{z} = \psi(\overline{x},\overline{y})$, then set $\phi(x,y) = \psi(x - x_0(\mu_0(y)),y) + z_0(\mu_0(y))$ and $z = \phi(x,y)$ will be required solution.

LEMMA 2: Solution of $C^m \to$ solution of $\overline{C}^m \equiv G^m \equiv I^m \to N^m$

$\overline{C}^m \to C^m$ provided $y_0'(\mu) \neq 0$ in M.

$N^m \rightarrow G^m$, provided $F_p \neq 0$ in U

$N^m \rightarrow \overline{C}^m$, provided $F_p - h'(y)\cdot F_q \neq 0$ in U

$N^m \rightarrow C^m$, provided $F_p \cdot y_0'[\mu_0(y)] - F_q \cdot x_0'[\mu_0(y)] \neq 0$ in U

and $y_0'(\mu) \neq 0$ in M.

Here $m = 1, 2, 3, \ldots$, or $m = +\infty$. [The solution of problem C^m usually is a function of class A^m in R, but that need not be required in problem. When it is so required, we shall indicate this by C^{m*}.]

LEMMA 3: Solution of $C_\delta \rightarrow$ Solution of $\overline{C}_\delta \equiv G_\delta \equiv I_\delta \rightarrow N_\delta \equiv IN_\delta$.

Solution of $\overline{C}_\delta \rightarrow$ Solution of C_δ, provided $|x_o'(\mu_o)| + |y_o'(\mu_o)| > 0$

Solution of $N_\delta \rightarrow$ Solution of G_δ, provided $F_p \neq 0$ at $p^0 \cdot (x^0, y^0, z^0, p^0, q^0)$

Solution of $N_\delta \rightarrow$ Solution of C_δ, provided

$$\Delta = \begin{vmatrix} x_0'(\mu_0) & y_0'(\mu) \\ F_p(p_0) & F_q(p_0) \end{vmatrix} \neq 0$$

Proof: Given a solution of $\overline{C}_\delta$ and given C_δ, with $|x_0'(\mu_0)| > 0$, either $x_0'(\mu_0) \neq 0$. If the latter is the case, then in some region T_δ about y_0, $y_0'(\mu) \neq 0$. Then apply lemma 2 to get result. If $x_0'(\mu_0) \neq 0$, we can use corresponding solution, when we interchange x and y in C_δ. Thus $\overline{C}_\delta \rightarrow C_\delta$.

The proof of the next statement is analogous. If $\Delta_0 \neq 0$, then either $x_0'(\mu_0) \neq 0$ or $y_0'(\mu_0) \neq 0$ and likewise either $F_p(x_0, ..) \neq 0$ or $F_q(x_0, ..) \neq 0$.

The generalization of the above Cauchy problems to the case of (n+1) independent variables - equation (6.3) - is made by replacing y by $(y_1, \ldots, y_n)$ and the interval T by a region T in E_n.

Linear equations:

$$(8.12) \qquad a(x,y)\frac{\partial z}{\partial x} + b(x,y)\frac{\partial z}{\partial y} + c(x,y)\,z + d(x,y) = 0$$

or quasi-linear equations:

$$(8.13) \qquad g(x,y,z)\cdot\frac{\partial z}{\partial x} + h(x,y,z)\frac{\partial z}{\partial y} + k(x,y,z) = 0.$$

have such special properties that theorems regarding the existence of their solutions will be given in separate sections.

9. In this section we shall describe certain geometric concepts usually used in discussing solutions of first order equations, (6.2), and define certain systems of ordinary differential equations, called characteristic equations, associated with (6.2). The results given here are not existence theorems, but are used in the statement and proof of those theorems.

Any five numbers D: (x_0,y_0,z_0,p_0,q_0) may be considered as defining a plane

$$(9.1) \qquad z - z_0 = p_0(x-x_0) + q_0(y-y_0)$$

in three-dimensional space which passes through (x_0,y_0,z_0) and has $(p_0:q_0:-1)$ as a set of direction numbers of its normal; conversely, every plane through (x_0,y_0,z_0) except a plane parallel to the z-axis will correspond to a set D. Hence we can call such a set D: (x_0,y_0,z_0,p_0,q_0) a *plane element* and consider that there is a family of such elements associated with every point (x_0,y_0,z_0) of space.

If $\phi(x,y)$ is a function of class A^1 in R, then the tangent plane to the surface S: $z = \phi(x,y)$ determines at each point of S a particular plane element:

$$(9.2) \qquad (x_0,y_0,z_0 = \phi(x_0,y_0),\ p_0 = \phi_x(x_0,y_0),\ q_0 = \phi_y(x_0,y_0))$$

which we call the *tangent element* of the surface at the point.

Given an equation

(9.3) $F(x,y,z,p,q) = 0$

all the plane elements at (x_0, y_0, z_0) which satisfy this equation. that is, for which $F(x_0, y_0, z_0, p_0, q_0) = 0$ correspond to tangent planes to a certain cone, called the elementary cone of (9.3). If a plane element is to be both the tangent element to a surface $z = \varphi(x,y)$ and a tangent plane of the elementary cone to (9.3) at a point (x_0, y_0, z_0), then from (9.2)

$$F(x_0, y_0, \varphi(x_0, y_0), \varphi_x(x_0, y_0), \varphi_y(x_0, y_0)) = 0$$

This means that $z = \varphi(x,y)$ must be a solution of

(6.2) $F(x,y,z,\frac{\partial z}{\partial x}, \frac{\partial z}{\partial y}) = 0.$

at the point (x_0, y_0, z_0). Hence, we call a plane element $(x_0, y_0, z_0, p_0, q_0)$ satisfying (9.3) an *integral element* of (6.2)

A space curve C, whose equations are: $x = x_0(\mu)$, $y = y_0(\mu)$, $z = z_0(\mu)$ where $x_0(\mu), y_0(\mu), z_0(\mu)$ are of class A^1 for $\mu \varepsilon M$, lies on a surface $z = \varphi(x,y)$ if

$$z_0(\mu) \equiv \varphi(x_0(\mu), y_0(\mu)) \qquad (\mu \varepsilon M_1 \leq M).$$

Hence

$$z_0'(\mu) \equiv \varphi_x(x_0(\mu), y_0(\mu)) \cdot x_0'(\mu) + \varphi_y(x_0(\mu), y_0(\mu)) \cdot y_0'(\mu).$$

At a point $[x_0 = x_0(\mu_0), y_0 = y_0(\mu_0), z_0 = z_0(\mu_0)]$ of C, the tangent element (9.2) of this surface must therefore satisfy:

$$z_0'(\mu_0) = p_0 \cdot x_0'(\mu_0) + q_0 \cdot y_0'(\mu_0).$$

As μ varies and C is generated, if we consider two functions $p(\mu)$ and $q(\mu)$, of class A^0 in M, which together with the equations of C should form a tangent element, then for all $\mu \varepsilon M$,

$$(9.4) \quad z_0'(\mu) = p_0(\mu) \cdot x_0'(\mu) + q_0(\mu) \cdot y_0'(\mu)$$

A set of functions:

$$(9.5) \quad x_0(\mu), y_0(\mu), z_0(\mu), p_0(\mu), q_0(\mu)$$

where the first three are of class A^1 and the last two of class A^0 in M, and such that (9.4) holds is called a *strip*. If a strip is an integral element of (6.2) at every point, that is, if it satisfies (9.3) so that

$$(9.6) \quad F(x_0(\mu), y_0(\mu), z_0(\mu), p_0(\mu), q_0(\mu)) \equiv 0$$

then it is called an *integral strip* of (6.2)

If a curve C is such that the tangent to C at each point is a ruling of the elementary cone of (9.3) at that point, it can be shown that $x'(\mu) : y'(\mu) : z'(\mu) = F_p : F_q : p\, F_p + q\, F_q$. If we choose the parameter so that $x'(\mu) = F_p$, $y'(\mu) = F_q$, then $z'(\mu) = p\, F_p + q\, F_q$; if $p(\mu)$ and $q(\mu)$ are chosen so that these three relations are satisfied, (9.4) will hold, and the five functions will form a strip. If this strip is to be at each point of C a tangent element to a surface $z = \varphi(x,y)$, which is a solution of (6.2) then since

$$F(x,y,\varphi(x,y),\varphi_x(x,y),\varphi_y(x,y)) = 0$$

$$F_x + F_z\, \varphi_x + F_p\, \varphi_{xx} + F_q\, \varphi_{yx} = 0$$

$$F_y + F_z \varphi_y + F_p \varphi_{xy} + F_q \varphi_{yy} = 0$$

and assuming $\varphi(x,y)$ is of class A^2, the second equation yields:

$$0 = F_x + F_z\, \varphi_x(x_0(\mu), y_0(\mu)) + x_0'(\mu) \cdot \varphi_{xx}(x_0(\mu), y_0(\mu))$$

$$+ y_0'(\mu) \cdot \varphi_{xy}(x_0(\mu), y_0(\mu))$$

$$= F_x + F_z \cdot p_0(\mu) + \frac{d}{d\mu}[\varphi_x(x_0(\mu), y_0(\mu)]$$

$$= F_x + F_z \cdot p_0(\mu) + p_0{}'(\mu).$$

Similarly

$$0 = F_y + F_z \cdot q_0(\mu) + q_0{}'(\mu).$$

Therefore we call a set of functions (8.5) which satisfy the equations:

$$(9.7) \quad \begin{cases} x' = F_p(x,y,z,p,q) \\ y' = F_q(x,y,z,p,q) \\ z' = pF_p(x,y,z,p,q) + qF_q(x,y,z,p,q) \\ p' = -F_x(x,y,z,p,q) - pF_z(x,y,z,p,q) \\ q' = -F_y(x,y,z,p,q) - qF_z(x,y,z,p,q) \end{cases}$$

a characteristic strip of (6.2) the equations (9.7) themselves are called the characteristic equations of (6.2) and form a set of ordinary differential equations in (μ,x,y,z,p,q) to which we can apply the theorems of Section 5.

THEOREM (9.1) Let $F(x,y,z,p,q)$ be a function of class A^1 in U; then along every characteristic strip (9.5) this function is constant. If a characteristic strip contains at least one integral element of (9.3) it is an integral strip of (6.2).

Proof: $\frac{d}{d\mu} F(x(\mu),y(\mu),z(\mu),p(\mu),q(\mu)) = F_x \cdot x' + F_y \cdot y' + F_z \cdot z' + F_p \cdot p'$
$+ F_q \cdot q' = F_xF_p + F_yF_q + F_z(pF_p + qF_q) + F_p(-F_x - pF_z)$
$+ F_q(-F_y - qF_z) = 0.$

$\therefore \quad F(x(\mu),y(\mu),z(\mu),p(\mu),q(\mu)) = C$, a constant, for all $\mu \varepsilon M$.

Now if this strip contains an integral element, there is some $\mu = \mu_0$ such that the corresponding plane element satisfies (9.3) Hence

$$F(x(\mu_0), z(\mu_0), p(\mu_0), q(\mu_0)) = 0$$

Thus $C = 0$ and hence for all $\mu \varepsilon M$,

$$F(x(\mu), y(\mu), z(\mu), p(\mu), q(\mu)) \equiv 0$$

and hence (9.5) satisfies (9.6) as well as (9.4) and is therefore an integral strip.

THEOREM 9.2: Let $F(x,y,z,p,q)$ be of class A^1 in a region U; let $z = \phi(x,y)$ be a solution of (6.2) which is of class A^2 in a region R. Then to every tangent element (9.2) of this surface there is at least one characteristic strip which contains this element for $\mu = 0$ and which lies on $z = \phi(x,y)$; this is an integral strip.

Proof: Let $(x_0, y_0, z_0, p_0, q_0)$ be given tangent element (9.2). There exists a solution $x = x_0(\mu)$, $y = y_0(\mu)$ of the equations:

$$x' = F_p(x,y,\phi(x,y),\phi_x(x,y),\phi_y(x,y))$$

$$y' = F_q(x,y,\phi(x,y),\phi_x(x,y),\phi_y(x,y))$$

which for $\mu = 0$ passes through (x_0, y_0). For some M_0, then, $(x_0(\mu), y_0(\mu)) \varepsilon R$, and M_0 contains $\mu = 0$. Now define

$$z_0(\mu) = \phi(x_0(\mu), y_0(\mu))$$

$$p_0(\mu) = \phi_x(x_0(\mu), y_0(\mu))$$

$$q_0(\mu) = \phi_y(x_0(\mu), y_0(\mu))$$

These five functions, for $\mu \varepsilon M_0$, define a strip which satisfies (9.7) and, for $\mu = 0$, they reduce to (9.2). For $\mu \varepsilon M_0$, they satisfy (9.6).

When $F(x,y,z,p,q) \equiv p - f(x,y,z,q)$, so that the equation is of normal type, the first equation of 9.7 is: $x' = 1$, so that if we take $x(\mu) = \mu$, , and then use x as a parameter instead of μ, equations (9.7) become:

$$y' = -f_q,\ z' = +p - qf_q,\ p' = +f_x + pf_z,$$

$$q' = +f_y + qf_z$$

But since on an integral strip the equation is satisfied: $p = f(x,y,z,q)$, we can substitute in the strip equation and get

$$(9.8) \qquad y' = f_q,\ z' = f - qf_q,\ q' = f_y + qf_z$$

as a set of equations from which $y_0(x)$, $z_0(x)$, $q_0(x)$ can be determined; then $p_0(x) = f(x,y(x),z(x),q(x))$ will satisfy the remaining equation $p^{(1)} = f_x + pf_z$ identically, and these together with $x_0(x) = x$, will form the characteristic strips. Hence (9.8) are often called the characteristic equations of the normal equation (8.5).

The characteristic equations of the general first order differential equation in (n+1) variables, (6.3), are defined in a manner analogous to (9.7) to be:

$$\begin{aligned}
x' &= F_p(x,y_1,\ldots y_n,z,p,q_1,\ldots q_n) \\
y_i' &= F_{q_i}(x,y_1,\ldots y_n,z,p,q_1,\ldots q_n) \qquad (i=1,\ldots,n) \\
(9.10) \qquad z' &= pF_p + \sum_{i=1}^{n} q_i F_{q_i} \\
p' &= -F_x - pF_z \\
q_i' &= -F_{y_i} - q_i F_z \qquad (i=1,\ldots,n)
\end{aligned}$$

B. The Initial Value Problem for Equations of Normal Type.

10. Cauchy[1], Darboux[1] and S. Kowalewski[1] were among the first to consider problem C; Kowalewski, particularly, studied problem N for a wide class of systems known as normal systems. Goursat and others simplified and clarified her results and proofs; we give the basic results below.

THEOREM 10.1: (a_∞) Let $g(y)$ be a function of class A^∞ for $|y-y_0|<\delta$ and let x_0 be a given number. Let $z_0 = g(y_0)$ and $q_0 = g'(y_0)$.

(b_∞) Let $f(x,y,z,q)$ be a function of class A^∞ in S_δ:

$$|x-x_0| < \delta, \quad |y-y_0| < \delta, \quad |q-q_0| < \delta.$$

Then there exists a unique function $\varphi(x,y)$ such that:

(α_∞) $\varphi(x,y)$ is of class A^∞ in some neighborhood R_{δ_1}:

$$|x-x_0| < \delta_1, \quad |y-y_0| < \delta_2 \text{ of } (x_0,y_0).$$

(β_0) For all $(x,y) \varepsilon R_\delta$, $z = \varphi(x,y)$ is a solution of:

$$(7.5) \quad \frac{\partial z}{\partial x} = f(x,y,z,\frac{\partial z}{\partial y})$$

(γ_0) For all $y \varepsilon T_{\delta_1}$ $|y-y_0| < \delta_1$

$$(10.5) \quad \varphi(x_0,y) \equiv g(y).$$

Proof: See Goursat[1], Horn[1], pp. 163-165.

By hypothesis:

$$(10.6) \quad f(x,y,z,q) = \sum_{\eta,\lambda,\mu,\nu=0}^{\infty} A_{\eta\lambda\mu\nu} (x-x_0)^\eta (y-y_0)^\lambda (z-z_0)^\mu (q-q_0)^\nu$$

where this series converges absolutely in S_δ. We wish to determine, first formally, the coefficients of

$$(10.7) \quad \varphi(x,y) = \sum_{\eta,\lambda=0}^{\infty} B_{\eta\lambda} (x-x_0)^\eta (y-y_0)^\lambda$$

to satisfy (β_0) and (γ_0) and then establish convergence in some R_{δ_1}. We also know that

$$(10.8) \quad g(y) = \sum_{\lambda=0}^{\infty} C_\lambda (y-y_0)^\lambda$$

where this converges absolutely in T_δ.

By Taylor's theorem

$$\varphi_{y\lambda} (x_0,y_0) = B_{0\lambda} \cdot \lambda! \text{ and } g^{(\lambda)}(y_0) = C_\lambda \cdot \lambda!$$

If (10.5) holds, then $\varphi_{y\lambda} (x_0,y) \equiv g^{(\lambda)}(y)$, and hence setting $y = y_0$ and substituting

$$B_{0\lambda} \cdot \lambda! = C_\lambda \cdot \lambda! \quad \text{or } B_{0\lambda} = C_\lambda \quad (\lambda = 0,1,2,\ldots).$$

In particular, $B_{0,0} = C_0 = z_0$ and $B_{01} = C_1 = q_0$.

If $z = \varphi(x,y)$ satisfies (8.6) in some R_{δ_1}, this must hold at (x_0, y_0):

$$\varphi_x(x_0,y_0) = f(x_0,y_0,\varphi(x_0,y_0),\varphi_y(x_0,y_0)).$$

Again from Taylor's theorem, right side $= f(x_0,y_0,z_0,q_0) = A_{0000}$ and left side $= B_{10}$, so

$$B_{10} = A_{0000}.$$

Differentiating the identity that would hold in R_δ:

$$(10.9)\quad \varphi_x(x,y) = f(x,y,\varphi(x,y),\varphi_y(x,y))$$

with respect to y and substituting (x_0,y_0), we get:

$$\varphi_{xy}(x_0,y_0) = f_y(x_0,y_0,z_0,q_0) + f_z(x_0,y_0,z_0,q_0)\cdot\varphi_y(x_0,y_0) + f_q(x_0,y_0,z_0,q_0)\cdot\varphi_{yy}(x_0,y_0).$$

Again substituting from (10.6) and (10.7) by Taylor's theorem:

$$B_{11} = A_{0100} + A_{0010}\,B_{01} + A_{0001}\,B_{02}\cdot 2.$$

Continuing this process, we establish by differentiating (10.9) λ times with respect to y and taking the result at (x_0,y_0), that every $B_{1\lambda}$ can be determined in terms of $A_{\eta\lambda\mu\nu}$'s, C_λ's and $B_{0\lambda}$'s, which again means in terms of $A_{\eta\lambda\mu\nu}$'s. Then if we differentiate $\varphi_{xy^\lambda}(x,y)$, as obtained from (10.9) η times with respect to x, and evaluate it at (x_0,y_0), we get $B_{\eta\lambda}$, again in terms of previous B's and A's and hence of $A_{\eta\lambda\mu\nu}$'s, and C_λ's. (Prove by finite induction.) Thus we have a method of calculating any coefficient $B_{\eta\lambda}$ of (10.7) and $B_{\eta\lambda}$ *is a rational integral function of* $A_{\eta\lambda\mu\nu}$*'s and* C_λ*'s with positive integers as coefficients.*

To establish convergence, the so-called *method of majorants* is used. If we can find a function $\bar{f}(x,y,z,q)$ of class A^{∞} in S_{δ} and a function $\bar{g}(y)$ of class A^{∞} in T_{δ}, so that the coefficients $\bar{A}_{\eta\lambda\mu\nu}$ and $\bar{C}_{\lambda}$ occurring in their power series are related to (10.6) and (10.7) by inequalities:

$$(10.10) \qquad |A_{\eta\lambda\mu\nu}| \leq \bar{A}_{\eta\lambda\mu\nu}; \qquad |C_{\lambda}| \leq \bar{C}_{\lambda}$$

and if we carry through the above process with regard to the equations

$$\frac{\partial z}{\partial x} = \bar{f}(x,y,z,q); \quad \bar{\varphi}(x_0,y) = \bar{g}(y)$$

getting a series solution $\bar{\varphi}(x,y) = \Sigma \bar{B}_{\eta\lambda}(x-x_0)^{\eta}(y-y_0)^{\lambda}$, then it follows from the italicized statement above that for all η and λ,

$$|B_{\eta\lambda}| \leq \bar{B}_{\eta\lambda}.$$

Hence if we can establish convergence of the series for $\bar{\varphi}(x,y)$ in some R_{δ_1}, then by the comparison test (9.7) converges absolutely in R_{δ_1}.

By Cauchy's inequality, if $M = \underset{S_\delta}{\text{l.u.b.}}\, f(x,y,z,q)$, then

$$(10.11) \qquad |A_{\eta\lambda\mu\nu}| \leq \frac{M}{\delta^{\eta}\delta^{\lambda}\delta^{\mu}\delta^{\nu}}.$$

Hence one could take:

$$\bar{f}(x,y,z,q) = \sum_{\eta,\lambda,\mu,\nu=0}^{\infty} \frac{M}{\delta^{\eta}\delta^{\lambda}\delta^{\mu}\delta^{\nu}} (x-x_0)^{\eta}(y-y_0)^{\lambda}(z-z_0)^{\mu}(q-q_0)^{\nu}$$

$$(10.12) \qquad = \frac{M}{\left(1-\frac{x-x_0}{\delta}\right)\left(1-\frac{y-y_0}{\delta}\right)\left(1-\frac{z-z_0}{\delta}\right)\left(1-\frac{q-q_0}{\delta}\right)}$$

and this certainly converges absolutely in S_{δ}.

But the terms of the above series are in turn less than those of another series, and so on, (Horn[1], Goursat[1], pp. 4-6) until one arrives at a series for a differential equation which can be reduced to any ordinary differential equation and whose solution exists in a certain R_δ. Thus we have convergence of the series for $\bar{\varphi}(x,y)$ and so for $\varphi(x,y)$. (This part of the proof usually carried out for problem IN and then transformation effected to get back to N, as in lemma 1.)

COROLLARY 1: The series in theorem (10.1) converges for at least $0 < |x-x_0| < r_1\delta$ and $0 < |y-y_1| < r_2\delta$ where $0 < r_1 < 1$ and $r_2 = \dfrac{(1-r_1)^2}{1+8M}$, M being defined in (10.11)

Proof: Perron[1]. He establishes this again when $x_0=y_0=z_0=q_0=0$ and $A_{0000}=0$, that is, for problem IN and where $\delta = 1$; showing successively that the series for (10.12) in this case is $\leq$ the series for:

$$\bar{\bar{f}}(x,y,z,q) = \frac{M}{(1-x)\left(1-\frac{y}{(1-x)^2}\right)\left(1-\frac{z}{1-x}-q\right)},$$

he solves:

$$\frac{\partial z}{\partial y} = \frac{M}{(1-x)\left(1-\frac{y}{(1-x)^2}\right)\left(1-\frac{z}{1-x}-\frac{\partial z}{\partial x}\right)}$$

by setting $u = \dfrac{y}{(1-x)^2}$ and $z = (1-x)\,F(u)$, getting:

$$F'(u) = \frac{1 \pm \sqrt{1-\frac{8u}{1-u}}}{4u} \quad \text{with } F(0) = 0 \text{ as condition.}$$

But since $F'(u)$ is regular at $u = 0$, this can be put into a series expansion, and hence

$$F'(u) = \sum_{n=0}^{\infty} \bar{B}_n u^n = M + (M + 2M^2)\,u + \ldots$$

Solution of above partial differential equation then reduces to:

$$z = (1-x)\int F'(u)\,du = (1-x)\int_0^{\frac{y}{(1-x)^2}} \frac{1-\sqrt{1-\frac{8M}{1-u}}}{4u}\,du = M\left(\frac{y}{1-x}\right)$$

$$+ \tfrac{1}{2}(M + 2M)^2 \frac{y^2}{(1-x)^3} + \ldots .$$

Then from determination of radius of convergence of series for $F^{(1)}(u)$, we see that the above converges for $|x| < 1$, $\left|\frac{y}{1-x}\right| < \frac{1}{1+8M}$ which leads to desired result.

Perron's proof is also interesting because in the formal determination of coefficients, he gets the B's from the A's by a formal scheme of comparing coefficients of like powers of y.

THEOREM 10.2

Let (a_∞) and (b_∞) be given as in theorem (10.1) Then there exists a uniformly convergent sequence of functions:

$$(10.13) \qquad \varphi^j(x,y) \qquad\qquad (j = 0,1,2,\ldots)$$

of class A^∞ in a neighborhood R_{δ_1} of (x_0, y_0) such that

$$(10.14) \qquad \lim_{j\to\infty} \varphi^j(x,y) = \varphi(x,y)$$

satisfies (α_∞), (β_0), (γ_0) and hence is the unique solution described in the previous theorem.

Proof: Germay[1], pp. 20. Germay does not use the method of majorants, but solves the characteristic equations of (8.5) (see Section 8)

$$(10.15) \quad \begin{cases} y' = -f_q(x,y,z,q) \\ z' = f(x,y,z,q) - q\, f_q(x,y,z,q) \\ q' = f_y(x,y,z,q) + q\, f_z(x,y,z,q) \end{cases}$$

by the method of successive approximation. By Section 5, if we take a set of values: $(x_0, \eta, g(\eta), g'(\eta))$ where η is any value in T_δ, we get a unique solution of the above system assuming these initial values for $x = x_0$:

$$y = y_0(x), \quad z = z_0(x), \quad q = q_0(x)$$

such that

$$y_0(x_0) = \eta; \; z_0(x_0) = g(\eta); \; q_0(x_0) = g'(\eta)$$

Doing this for any η in T_δ, we can write the solutions

$$(10.16) \quad y = Y(x,\eta); \; z = Z(x,\eta); \; q = \varphi(x,\eta).$$

(The general characteristic functions on page 10 would be $y = Y(x, x_0, \eta)$, etc., but here x_0 is fixed, throughout the problem.) These are limits of $Y^j(x,\eta), \ldots$; as in theorem (6.4) $Y_\eta > 0$, and hence the first equation of (10.16) can be solved, by the implicit function theorem, for $\eta = \theta(x,y)$ and this solution substituted in the second to get $z = Z(x,\theta(x,y))$. Taking $\varphi(x,y) \equiv Z(x,\theta(x,y))$, this function is shown to satisfy (α_∞), (β_0), (γ_0).

This method had been used previously; Germay's new idea was to carry through the same process for the approximating functions $Y^j(x,\eta)$ and $Z^j(x,\eta)$ and to show that the inverse $\theta^j(x,\eta)$ converges uniformly to $\theta(x,y)$, so the sequence $\varphi^j(x,y) = Z^j(x,\theta^j(x,y))$ $(j=0,1,2,\ldots)$ will converge uniformly to $\varphi(x,y)$. The above theorems can be extended to n+1 variables:

THEOREM 10.3 *

(a_∞) Let $(x^0, y_1^0, \ldots, y_n^0)$ be a given point and let $g(y_1, \ldots, y_n)$ be a function of class A^∞ in a neighborhood T_δ of $(y_1^0, \ldots, y_n^0)$. Let $z_0 = g(y_1^0, \ldots, y_n^0)$ and $q_k^0 = g_{y_k}(y_1^0, \ldots, y_n^0)$. $(k=1,2,\ldots n)$

(b_∞) Let $f(x, y_1, \ldots, y_n, z, q_1, \ldots, q_n)$ be a function of class A^∞ in a neighborhood S_δ of $(x_0, y_1^0, \ldots, y_n^0, z^0, q^0, \ldots, q_n^0)$.

Then there exists a unique function $\varphi(x, y_1, \ldots, y_n)$ such that

(α_∞) $\phi(x,y_1,\ldots,y_n)$ is of class A^∞ in a neighborhood R_δ of $(x^0,y_1^0,\ldots,y_n^0)$

(β_0) $z = \phi(x,y_1,\ldots,y_n)$ is a solution in R_{δ_1} of:

$$(10.2)\quad \frac{\partial z}{\partial x} = f\left(x,y_1,\ldots,y_n,z,\frac{\partial z}{\partial y_1},\ldots,\frac{\partial z}{\partial y_n}\right)$$

(γ_0) $\phi(x_0,y_1,\ldots,y_n) \equiv g(y_1,\ldots,y_n)$ in T_{δ_1}.

Proof: Goursat[1], pp. 2-6. See Perron[1] for corresponding corollary.

THEOREM 10.4: Let (a_∞) and (b_∞) be given as above. Then there exists a uniformly convergent sequence of functions

$$\phi^j(x,y_1,\ldots,y_n) \qquad (j = 0,1,2\ldots)$$

of class A^∞ in a neighborhood of $(x_0,y_1^0,\ldots,y_n^0)$ such that

$$\lim_{j\to\infty} \phi^j(x,y_1,\ldots,y_n) = \phi(x,y_1,\ldots,y_n)$$

satisfies (α_∞), (β_0), (γ_0), and hence is the function of theorem 10.1.

Proof: Germay[1] p. 20

11. The method outlined in the first part of the proof of theorem 10.2 - the so-called method of characteristics - does not depend upon the power series expansions of f or g, provided one uses theorems which do not depend upon power series expansions for the corresponding ordinary differential equations. Consequently later writers have made this the basis of existence theorems which apply when the requirement that $\phi(x,y)$ be analytic is replaced by the requirement that it be of class A^2. At the same time, solutions were obtained which held in the large - that is, in a given region R, or a specified subregion of R. Results are often given for regions of infinite extent in y, and then results for finite regions derived as corollaries. For this reason, the hypotheses in an infinite region may carry additional requirements concerning boundedness.

THEOREM 11.1: (a_2) Let x_0 be a given number and let $g(y)$ be of class A^2 for all y.

(b_2) Let $f(x,y,z,q)$ be of class A^2 in:

S_1: $|x-x_0| < a$, $-\infty<y<+\infty$; $-\infty<z<+\infty$; $-\infty<q<+\infty$.

(c) Let there be a constant B such that $1 + |g'(y)| + |g(y)| \leq B$ for all y, and a constant A such that

$$|f| < A, \quad |f_z| < A, \quad |f_y| < A, \quad |f_q| < A, \quad |f_{yy}| < A,$$

(11.1) $$|f_{yq}| < A, \quad |f_{yz}| < A, \quad |f_{qq}| < A, \quad |f_{zz}| < A, \quad |f_{qz}| < A.$$

Let $\alpha = \min\left(a, \frac{1}{3AB}\right)$.

Then there exists a unique function $\varphi(x,y)$ such that:

(α_2) $\varphi(x,y)$ is of class A^2 everywhere in R_1: $|x-x_0| < \alpha$, $-\infty < y < +\infty$.

(β_0) $z = \varphi(x,y)$ is a solution in R of

(8.5) $\frac{\partial z}{\partial x} = f(x,y,z,\frac{\partial z}{\partial y})$

(γ_0) For all y, $\varphi(x_0,y) \equiv g(y)$.

Proof: Kamke[1], pp. 352-358. As in the proof of theorem 10.2 pp. 37 the characteristic equations (10.14) of (8.5) are solved, for any initial value η, and the corresponding $(x_0, \eta, g(\eta), g'(\eta))$ to get

(10.16) $y = Y(x,\eta)$, $z = Z(x,\eta)$, $q = Q(x,\eta)$

Then it is shown that $Y_\eta(\eta) > 0$ for all y and hence that first function has a unique inverse $\eta = \theta(x,y)$. The function $\varphi(x,y) = Z(x,\theta(x,y))$ then turns out to have properties (α_2), (β_2), (γ_0).

(The method of obtaining (10.16) is amenable to computation, but finding the inverse may not be. It may be that Germay's work has possibilities if extended to this case.) The above result was somewhat improved by Ważewski[1] and Digel[1].

THEOREM 11.2 .

(a_1) Let x_0 be a given number and let g(y) be a function of class A^1 for all y.

(b_1) Let $f(x,t,z,q)$ f_x, f_y, f_z be of class A^0 in S:

$|x-x_0| < a;\ -\infty<y,z,q<+\infty.$

($\bar{c}_1$) Let $|f| < A$, $|f_y| < A$, $|f_z| < A$, $|f_q| < A$ in S_1 and let g, f, f_y, f_z, f_q satisfy Lipschitz conditions with the same constant M:

$$|g(y)-g(\bar{y})| \leq M\ |y-\bar{y}|$$

$$|f(x,y,z,q)-f(x,\bar{y},\bar{z},\bar{q})| \leq M[\,|y-\bar{y}| + |z-\bar{z}| + |q-\bar{q}|\,]$$

$$|f_y(x,y,z,q)-f_y(x,\bar{y},\bar{z},\bar{q})| \leq \quad ''$$

$$|f_z(x,y,z,q)-f_z(x,\bar{y},\bar{z},\bar{q})| \leq \quad ''$$

$$|f_q(x,y,z,q)-f_q(x,\bar{y},\bar{z},\bar{q})| \leq \quad ''$$

Set $\alpha = \min\left(a, \frac{1}{4AB}\right)$; assume $1 + |g'(y)| + M \leq B$.

Then there exists a function $\varphi(x,y)$ such that

(α_1) $\varphi(x,y)$ is of class A^1 in R_1: $|x-x_0| < \alpha$, $-\infty < y < +\infty$

(β_0) For (x,y) in R_1, $z = \varphi(x,y)$ is a solution of (8.5)

$$\frac{\partial z}{\partial x} = f\left(x,y,z,\frac{\partial z}{\partial y}\right)$$

(γ_0) For all y, $\varphi(x_0,y) \equiv g(y)$.

Proof: Ważewski[1] proved this by approximating polynomials. Digel[1] used a proof similar to the one given by Kamke for the previous theorem; it appears neater than Ważewski's, but the latter may be more practical for computation.

Finally, we have a result established by Kamke[4] in 1943 as part of a general theorem involving solutions with respect to parameters in n variables. Note that if $g(y)$ and $f(x,y,z,q)$ are of class A^{m+1}, for any $m \geq 1$, the requirement that $\varphi(x,y)$ be of class A^2 is enough to produce a unique solution; as a consequence, this solution will turn out to be of class A^{m+1}.

THEOREM 11.3

(a_{m+1}) Let x_0 be a given number and let $g(y)$ be of class A^{m+1} ($m \geq 1$) for all y.

(b_{m+1}) Let $f(x,y,z,q)$ be of class A^{m+1} or, more generally, let f, f_y, f_z, f_q be of class A^m in S_2: $|x-x_0| \leq a$, $-\infty<y,z,q < +\infty$.

($\bar{c}$) Let there be a constant B such that $|g'(y)| + |g''(y)| \leq B$ for all y, and a constant A such that (11.1) (p. 40) holds in S_2.

Set $0 < \beta < \frac{1}{A} \log\left(1 + \frac{3}{2(B+1)}\right)$ and $\alpha = \min(a,\beta)$.

Then there is a unique function $\varphi(x,y)$ such that:

(α_2) $\varphi(x,y)$ is of class A^2 in R_2: $|x-x_0| \leq \alpha$, $-\infty<y<+\infty$

(β_0) $z = \varphi(x,y)$ is a solution in R_2 of (7.5):

$$\frac{\partial z}{\partial x} = f\left(x,y,z,\frac{\partial z}{\partial y}\right)$$

(γ_0) $\varphi(x_0,y) \equiv g(y)$ for all y.

Furthermore, $\varphi(x,y)$ is of class A^{m+1} in R_2.

COROLLARY 1: If S_2 is replaced by S_1: $|x-x_0| < a$, $-\infty<y,z,q<+\infty$, in condition (b_{m+1}) then R_2 is replaced by R_1: $|x-x_0| < \alpha$, $-\infty<y<+\infty$ in (α_2).

COROLLARY 2: We can take $\beta = \frac{1}{3(B+1)}$ so that $\alpha = \min\left(\frac{\log 3}{2(B+1)}, a\right)$.

Both theorem 11.1 and theorem 11.3 can be applied when the given regions for $g(y)$ and $f(x,y,z,q)$ are finite. This is done by using the following lemma, which we give for (n+1) dimensions:

LEMMA: Let $f(x,y_1,\ldots,y_n)$ be continuous and have continuous first and second derivatives with respect to $y_1,\ldots,y_n$ in the region:

$$R_0: \quad a<x<b, \; c_1<y<d_1, \ldots, c_n<y_n<d_n$$

Suppose that $|f| \leq A$, $|f_{y_k}| \leq B$, $|f_{y_\nu y_\mu}| \leq C$. Then for every δ with $0<\delta< \underset{\nu}{\mathrm{Min}} \frac{d_\nu - c_\nu}{2}$, there exists a function $\bar{f}(x,y_1,\ldots,y_n)$ defined in:

$$R_1: \quad a<x<b, \; -\infty<y_1<+\infty, \ldots, -\infty<y_n<+\infty$$

which is continuous, has continuous first and second partial derivatives with respect to $y_1,\ldots,y_n$, in R_1 and $|\bar{f}| \leq A$, $|\bar{f}_{y_k}| \leq B$; $|\bar{f}_{y_\nu y_\mu}| \leq C + \frac{B}{\delta}$;

$(x,y_1,\ldots,y_n)$ in R_0, $\bar{f}(x,y_1,\ldots,y_n) = f(x,y_1,\ldots,y_n)$.

The theorems given above are true for (n+1) dimensions. Theorem (11.3) becomes:

THEOREM 11.4

(a_{m+1}) Let x_0 be a given number with $|x-x_0| \leq a$, and let $g(y_1,\ldots,y_n)$ be of class A^{m+1} for all y.

(b_{m+1}) Let $f(x,y_1,\ldots,y_n z,q_1,\ldots,q_n)$ be of class A^{m+1}, or more generally, let f, f_{y_k}, f_z, f_{q_k} be of class A^m in S_2:

$$|x-x_0| \leq a, \; -\infty<y_1,\ldots,y_n,z,q_1,\ldots,q_n<+\infty.$$

(c) Let there be a constant B such that

$$|g_{y_k}| + \sum_{j=1}^{k} |g_{y_k y_j}| \leq B \quad (k=1,\ldots,n)$$

and a constant A such that

$$|f_x|, \; |f_{y_k}|, \; |f_z|, \; |f_{q_k}|, \; |f_{y_k y_j}|, \; |f_{y_k z}|, \; |f_{y_k q}|,$$
$$|f_{zz}|, \; |f_{z q_k}|, \; |f_{q_j q_k}| \text{ are all } \leq A.$$

Set $0<\beta< \frac{1}{A} \log \left(1 + \frac{\log 3}{2n(B+1)}\right)$ and $\alpha = \min(a,\beta)$.

Then there exists a unique function $\varphi(x,y_1,\ldots,y_n)$ such that

(α_2) $\varphi(x,y_1,\ldots,y_n)$ is of class A^2 in R_2:

$$|x-x_0| \leq \alpha, \quad -\infty<y_1,\ldots,y_n<+\infty$$

(β_0) $z = \varphi(x,y_1,\ldots,y_n)$ is a solution in R_2 of

$$(10.2) \quad \frac{\partial z}{\partial x} = f(x,y_1,\ldots,y_n,z,\frac{\partial z}{\partial y_1},\ldots,\frac{\partial z}{\partial y_n})$$

(γ_0) $\varphi(x_0,y_1,\ldots,y_n) = g(y_1,\ldots,y_n)$ for all y.

Furthermore, $\varphi(x,y_1,\ldots,y_n)$ is of class A^{m} in R_2.

COROLLARY 1 If S_2 is replaced by S_1: $|x-x_0| <a$, $-\infty<y_1,\ldots,y_n<+\infty$, this result is valid when R_2 is replaced by R_1: $|x-x_0| < \alpha$, $-\infty<y_1,\ldots,y_n<+\infty$.

Recently, Baiada[1] has given a proof of theorem 11.2 which is interesting because it is a generalization of the Peano method in ordinary differential equations (Th.5.2). He uses the interval $0\leq x\leq a$ instead of $|x-x_0|\leq\delta$, and $x_0=0$.

In problem N, the initial line $x = x_0$ is taken as interior to the region R in which the solution $\varphi(x,y)$ exists; since φ is continuous in R, $\varphi(x,y) \to g(y)$, as $(x,y) \to (x_0,y)$ in any manner whatever. The situation is not essentially altered if R is replaced by a closed region $\bar{R}$ which has the line $x=x_0$ as part of its boundary. However, a different kind of problem is obtained if one asks for a solution of class A^1 in the interior of a region R, which has this line as part of the boundary, and such that $\lim_{x\to x_0} \varphi(x,y) = g(y)$. For such a solution, if it exists, need not be continuous at (x_0,y). Hadamard[1] pointed out this distinction for second order equations, but I have never seen a solution of this problem (call it problem H) explicitly stated and proved for first order equations.

C. *Solution of Problem of Cauchy and Initial Value Problem; Equation Not in Normal Form.*

12. Solution of Problems G and I

From the lemmas of section 8, the existence theorems of section B can be extended when the equation is not in normal form, provided certain additional conditions hold concerning $F(x,y,z,p,q)$ or the curve Γ.

Theorem 12.1

(a_∞) Let (x_0, y_0) be a given point. Let $g(y)$ be a function of class A^∞ for $|y-y_0| < \delta$. Set $z_0 = g(y_0)$ and $q_0 = g'(y_0)$.

(b_∞) Let $F(x,y,z,p,q)$ be a function of class A^∞ in a neighborhood U_δ: $|x-x_0| < \delta$, $|y-y_0| < \delta$, $|z-z_0| < \delta$, $|p-p_0| < \delta$, $|q-q_0| < \delta$ where p_0 is chosen so that:

$$(12.1) \quad \begin{aligned} &F(x_0, y_0, z_0, p_0, q_0) = 0 \\ &F_p(x_0, y_0, z_0, p_0, q_0) \neq 0 \end{aligned}$$

Then there exists a unique function $\varphi(x,y)$ such that:

(α_∞) $\varphi(x,y)$ is a function of class A^∞ in R_{δ_1}: $|x-x_0| < \delta_1$, $|y-y_0| < \delta_1$

(β_0) For all $(x,y) \varepsilon R_{\delta_1}$, $z = \varphi(x,y)$ is a solution of

$$(6.2) \quad F(x,y,z\, \frac{\partial z}{\partial x}, \frac{\partial z}{\partial y}) = 0$$

(γ_0) For all $y \varepsilon T_{\delta_1}$: $|y-y_0| < \delta_1$, $\varphi(x_0,y) \equiv g(y)$.

Proof: (See Goursat[1], p. 45, Horn[1], p. 162, ff.) As on page 24 solve $F(x,y,z,p,q) = 0$ for $p = f(x,y,z,q)$, and find solution of corresponding problem N with same (γ_0). This solution $\varphi(x,y)$, found as in theorem 10.1 will be required solution (see page 24).

THEOREM 12.2

The function $\varphi(x,y)$ of the previous theorem can be uniformly approximated by a convergent sequence: (10.13) $\{\varphi^j(x,y)\}$ $j=0, 1, 2, \ldots$ Furthermore, the derivatives of $\varphi^j(x,y)$ will converge uniformly to the derivatives of $\varphi(x,y)$.

Proof: Germay[1], p. 28. (See theorem 10.2 for a possibly useful procedure.)

Problem I^∞ is a special case of C and can be handled the same way.

To solve problem $\overline{C}^\infty$, let y_0 be given, set $x_0 = h(y_0)$ and $z_0 = g(y_0)$ and find p_0 and q_0 so that

$$F(x_0, y_0, z_0, p_0, q_0) = 0$$

$$g_0'(y_0) = h'(y_0) \cdot p_0 + q_0$$

If one assumes that

$$(12.2) \quad F_p(x_0, y_0, z_0, p_0, q_0) - h'(y_0) \cdot F_q(x_0, y_0, z_0, p_0, q_0) \neq 0$$

then $F_p - h'(y)F_q \neq 0$ in a neighborhood of $P_0(x_0, y_0, z_0, p_0, q_0)$ and hence by lemma 3 of section 8 the problem can be reduced to problem IN, by the formulas on page 25. Thus one establishes the existence of a solution in a neighborhood of P_0.

To solve problem C^∞, let $\mu_0 \varepsilon M$, set $x_0 = x_0(\mu_0), y_0 = y_0(\mu_0), z_0 = z_0(\mu_0)$, and find p_0 and q_0 so that

$$F(x_0, y_0, z_0, p_0, q_0) = 0$$

$$z_0'(\mu_0) = x_0'(\mu_0) \cdot p_0 + y_0'(\mu_0) \cdot q_0$$

If one assumes that

$$(12.3) \quad F_p(x_0, y_0, z_0, p_0, q_0) \cdot y_0'(\mu_0) - F_q(x_0, y_0, z_0, p_0, q_0) \cdot x_0'(\mu_0)$$

$$F_p(x, y. z, p, q) \cdot x_0'(\mu_0(y)) - F_q(x, y, z, p, q) \cdot y_0'(\mu_0(y)) \neq 0$$

in a neighborhood of P^0 and hence by the lemmas of section 8, reducing the problem by the formulas on page 25, existence of a solution of C^∞ is proved.

If one is no longer interested in analytic solutions, the same conditions (12.1), (12.2) or (12.3) would insure the existence of a solution of problem G_δ, $\overline{C}_\delta$, C_δ in the neighborhood of P^0, provided that after the transformations on page 25 were performed, the reduced problem N or IN satisfied the hypotheses of one of the theorems of section 11. Indeed, one could get a solution in the large of these problems, provided that, throughout a region U,

$$(12.4) \quad F_p(x, y, z, p, q) \neq 0$$

for problem G, or

$$(12.5) \quad F_p - h'(y) \cdot F_q \neq 0$$

when problem $\overline{C}$ is given, or

(12.6) $\quad y_0'(\mu) \neq 0;\ F_p \cdot y_0'(\mu_0(y)) - F_q \cdot x_0'(\mu_0(y)) \neq 0$

(12.7) $\quad x_0'(\mu) \neq 0;\ F_p \cdot y_0'(\mu_0(x)) - F_q \cdot x_0'(\mu_0(x)) \neq 0$

when problem C is given, where $\mu_0(y)$ and $\mu_0(x)$ are the inverses of $y_0(\mu)$ and $x_0(\mu)$. Thus, given problem C and (12.6), find the inverse $\mu = \mu_0(y)$ of $y = y_0(\mu)$, and determine $g(y) = z_0(\mu_0(y))$ and $h(y) = x_0(\mu_0(y))$. Then let:

(12.8) $\quad \overline{F}(\overline{x},\overline{y},\overline{z},\overline{p},\overline{q}) \equiv F(\overline{x}+g(\overline{y}),\ \overline{y},\ \overline{z}+h(\overline{y}),\ \overline{p},\ \overline{q} - h'(\overline{y})\overline{p})$

where $h'(\overline{y}) = x_0'(\mu_0(\overline{y})) \cdot \mu_0'(\overline{y})$.

Solve $\overline{F}(\overline{x},\overline{y},\overline{z},\overline{p},\overline{q}) = 0$ for $\overline{p} = \overline{f}(\overline{x},\overline{y},\overline{z},\overline{q})$, which is possible by (12.6), which becomes $\overline{F}_{\overline{p}}(\overline{x},\overline{y},\overline{z},\overline{p},\overline{q}) \neq 0$ for some region. Then find the solution of

$$\frac{\partial \overline{z}}{\partial x} = \overline{f}(\overline{x},\overline{y},\overline{z},\frac{\partial \overline{z}}{\partial y})$$

with boundary condition

$$\psi(0,\overline{y}) = 0.$$

This is problem IN, and so methods of section 11 are applicable. If the solution is $\overline{z} = \psi(\overline{x},\overline{y})$, then solution of problem C is $z = \phi(x,y)$ where

$$\phi(x,y) \equiv \psi(x - x_0(\mu_0(y)),\ y) + z_0(\mu_0(y)).$$

13. When the above conditions do not hold throughout a region, it is still possible to find a solution of problem C in the " neighborhood of the curve Γ". Thus, if $F(x,y,z,p,q)$ is of class A^2 in a region U and if $x_0(\mu), y_0(\mu), z_0(\mu)$ are functions of class A^1, for μ in M, so that Γ: $x=x(\mu), y=y(\mu), z=z(\mu)$ lies in the projection of U in (x,y,z) space, let

(13.1) $$\Delta = \begin{vmatrix} x_0'(\mu) & y_0'(\mu) \\ F_p(x_0(\mu),y_0(\mu),z_0(\mu),p,q) & F_q(x_0(\mu),y_0(\mu),z_0(\mu),p,q) \end{vmatrix} \neq 0$$

for μ in M and (p,q) in V. Then by the implicit function theorem, one can find solutions $p = p(\mu)$, $q = q(\mu)$ of the equations:

(13.1a) $$\begin{aligned} p \cdot x_0'(\mu) + q \cdot y_0'(\mu) &= z_0'(\mu) \\ F(x_0(\mu),y_0(\mu),z_0(\mu),p,q) &= 0 \end{aligned}$$

The functions

$$(13.2) \qquad [x_0(\mu), y_0(\mu), z_0(\mu), p_0(\mu), q_0(\mu)] \qquad (\mu \varepsilon M_1)$$

will form an initial integral strip (see page 29) of the equation

$$(6.2) \qquad F(x, y, z, \frac{\partial z}{\partial x}, \frac{\partial z}{\partial y}) = 0$$

Furthermore

$$(13.3) \qquad \Delta(\mu) = \begin{vmatrix} x_0'(\mu) & y_0'(\mu) \\ F_p(x_0'(\mu), \ldots, q_0'(\mu)) & F_q(x_0'(\mu), \ldots, z_0'(\mu)) \end{vmatrix} \neq 0 \qquad (\mu \varepsilon M_1)$$

Either $x_0'(\mu) \neq 0$ or $y_0'(\mu) \neq 0$, so that in the neighborhood of every point of the strip, either (12.6) or (12.7) holds; by carrying through the transformation indicated there, one establishes the existence of a solution in a region R containing $[x_0(\mu), y_0(\mu)]$. However, instead of working with the characteristic equations of the transformed normal equation, one can obtain the solution directly as follows: The characteristic equations:

$$(9.7) \qquad \begin{cases} x' = F_p(x, y, z, p, q) \\ y' = F_q(x, y, z, p, q) \\ z' = pF_p + qF_q \\ p' = -F_x - pF_z \\ q' = -F_y - qF_z \end{cases}$$

are solved, in (t,x,y,z,p,q) - space, for each set of initial values (13.2) to give

$$(13.4) \qquad \begin{cases} x = X(t; x_0(\mu), y_0(\mu), z_0(\mu), p_0(\mu), q_0(\mu)) \equiv X(y, \mu) \\ y = Y(t, \mu) \\ z = Z(t, \mu) \\ p = P(t, \mu) \\ q = Q(t, \mu) \end{cases}$$

By theorem (9.1) these will form an integral strip; as μ varies, the first three of these equations parametrically determine a surface $z = \varphi(x, y)$ which is a solution of (6.2) and passes through Γ, since one can show that $X_t Y_\mu - Y_t X_\mu \neq 0$.

Solving the first two equations, and substituting in the third, one obtains its explicit formula. This method does not give the region of existence of the solution.

Sometimes the problem of Cauchy is stated as that of determining a solution of (6.2), not through a given curve Γ, but through a given initial strip (13.2). Then one can be sure of the uniqueness of the solution for that strip (in the case of a given curve, uniqueness can be established provided one always has a unique solution of (13.1a)). Thus, Courant-Hilbert[1], v. 2, pp.66-68, proves:

THEOREM 13.1: (a) Let $x_0'(\mu), y_0(\mu), z_0(\mu)$ be functions of class A^1 with $x_0'^2+y_0'^2>0$.

(b) Let $F(x,y,z,p,q)$ be of class A^2 in U with $F_p^2+F_q^2>0$

(c) Let $p_0(\mu), q_0(\mu)$ be 2 additional functions determined so that (13.2) is an integral strip of (6.2) and (13.3) holds.

Then there exists a function $\phi(x,y)$ such that

(α_2) $\phi(x,y)$ is of class A^2 in a region R containing $(x_0(\mu), y_0(\mu))$

(β_0) $z = \phi(x,y)$ satisfies (6.2) in R.

(γ) $z_0(\mu) = \phi(x_0(\mu), y_0(\mu))$; $p_0(\mu) = \phi_x(x_0(\mu), y_0(\mu))$; $q_0(\mu) = \phi_y(x_0(\mu), y_0(\mu))$

For a given integral strip (13.2) satisfying (13.3), the solution is unique. Again, the region R is not defined precisely.

(Condition (13.3) means that at no point is the initial strip tangent to any characteristic strip (9.7) of the equation, since along a characteristic strip $F_p \cdot y' - F_q \cdot x' \equiv 0$. If the initial strip is a characteristic strip, there is no unique solution to the problem. See Courant-Hilbert, p. 69.)

Gross[1], and others, widen the concept of a solution of (6.2) by considering it to be a 2-dimensional manifold in (x,y,z,p,q) space, such as (13.4) which is the solution of (6.2) and the equations:

$$\frac{\partial z}{\partial u} = p\frac{\partial x}{\partial u} + q\frac{\partial y}{\partial u}, \quad \frac{\partial z}{\partial v} = p\frac{\partial x}{\partial v} + q\frac{\partial y}{\partial v}$$

He obtains existence theorems, and discusses singularities, but such generalized solutions are not within the scope of this paper.

Under the assumptions that solutions exist, there are other methods of obtaining them - such as the Lagrange method of complete integrals (see Courant-Hilbert) or the method of first integrals (Courant-Hilbert or Kamke[1], pp. 362-373) but these are not existence theorems and so are not included here.

For n+1 independent variables, the statement and solution of problems G, I, $\overline{C}$ are direct generalizations of the problems for 2 independent variables, where $z = g(y)$ is replaced by $z = g(y_1, \ldots, y_n)$ and, in problem $\overline{C}$, $x = h(y)$ is replaced by $x = h(y_1, \ldots, y_n)$, (see Goursat[1], and Germay[1], for analytic solutions.) In problem C, the generalization of the initial curve Γ in terms of a parameter μ, is an n-dimensional manifold

$$(13.5) \qquad x = \overset{o}{x}(\mu_1, \ldots, \mu_n), \; y_i = \overset{o}{y_i}(\mu_o, \ldots, \mu_n), \; z = z(\mu_1, \ldots, \mu_n)$$

where

$$J = \left\| \frac{\partial \overset{o}{x}}{\partial \mu_k} \quad \frac{\partial \overset{o}{y_1}}{\partial \mu_k} \cdots \frac{\partial \overset{o}{y_n}}{\partial \mu_k} \right\|$$

is of rank n. If the determinant $\left| \frac{\partial \overset{o}{y_i}}{\partial \mu_k} \right| \neq 0$, one can solve n of the equations (13.5 for $\mu_1, \ldots, \mu_n$ and substitute in the other 2, and use the methods of section 8 to get the corresponding problem $\overline{C}$; then investigation of the function F will determine whether or not the problem can be transformed to problem IN, and the theorems of section 11 applied. Or one can get a strip manifold Γ, from (13.5) by adding functions $\overset{o}{p}, \overset{o}{q_1}, \ldots, \overset{o}{q_n}$ of $\mu_o, \ldots, \mu_n$ such that

$$F[\overset{o}{x}, \overset{o}{y_1}, \ldots, \overset{o}{y_n}, \overset{o}{z}, \overset{o}{p}, \overset{o}{q_1}, \ldots, \overset{o}{q_n}) \equiv 0 \text{ in } (\mu_1, \ldots \mu_n)$$

$$\frac{\partial \overset{o}{z}}{\partial \mu_k} \equiv \overset{o}{p} \cdot \frac{\partial \overset{o}{x}}{\partial \mu_k} + \sum_{i=1}^{n} \overset{o}{q_i} \cdot \frac{\partial y_i}{\partial \mu_k} \qquad (k=1, \ldots, n)$$

If F is a function of class A^2 in a 2n+3 dimensional region U, containing a strip manifold Γ, of the above type, where all functions involved are of class A^1 and where

$$(13.6) \quad \Delta = \begin{vmatrix} F_p & F_{q_1} & \cdots & F_{q_n} \\ \frac{\partial x_o}{\partial \mu_1} & \frac{\partial y_1}{\partial \mu_1} & \cdots & \frac{\partial y_n}{\partial \mu_1} \\ \cdot & \cdot & \cdots & \cdot \\ \frac{\partial x_o}{\partial \mu_n} & \frac{\partial y_1}{\partial \mu_n} & \cdots & \frac{\partial y_n}{\partial \mu_n} \end{vmatrix} \neq 0 \text{ on } \Gamma_1 \, ,$$

there exists a solution of

$$F(x, y_1, \ldots, y_n, z, \frac{\partial z}{\partial x}, \frac{\partial z}{\partial y_1}, \ldots, \frac{\partial z}{\partial y_n}) = 0$$

in a neighborhood of $[x_0(\mu_1, \ldots \mu_n), \ldots, y_n(\mu_1, \ldots, \mu_n)]$ which passes through Γ. (Courant-Hilbert, v.2, p. 86.) The solution is unique for a given strip Γ_1. (As in 2 dimensions, if $\Delta \equiv 0$ on Γ_1, then Γ_1 must be a characteristic manifold in order that there be a solution of the problem, and in that case, there are an infinite number of solutions.

D. Linear and Quasi-linear Equations

14. Quasi-linear Equations - General Properties.

For the quasi-linear equation

$$a(x,y,z) \frac{\partial z}{\partial x} + b(x,y,z) \frac{\partial z}{\partial y} + c(x,y,z) = 0$$

the characteristic strip equations can be simplified, as they were for the normal form; hence one applies the term characteristic equations to:

$$x' = a(x,y,z)$$
$$y' = b(x,y,z)$$
$$z' = c(x,y,z)$$

These equations have special properties, which we shall discuss for n+1 dimensions, since they are readily comprehended.

Let $f^{\nu}(y_1, \ldots, y_n, z)$ $(\nu = 0,1,2,\ldots,n)$ be continuous in a region U of (n+1) dimensional space, and form the quasi-linear equation:

$$(14.1) \qquad \sum_{\nu=1}^{n} f^{\nu}(y_1, \ldots, y_n, z) \cdot \frac{\partial z}{\partial y_\nu} = f^{0}(y_1, \ldots, y_n, z)$$

The characteristic equations of (14.1) are the (n+1) ordinary differential equations, in the (n+2) dimensional space of $(t, y_1, \ldots, y_n, z)$:

$$(14.2) \qquad y_\nu' = f^{\nu}(y_1, \ldots, y_n, z)$$
$$z' = f^{0}(y_1, \ldots, y_n, z)$$

Each solution of this system:

$$(14.3) \qquad y_1 = \phi_1(t), \ldots, y_n = \phi_n(t), \; z = \phi_0(t)$$

is called a characteristic curve of (14.1)

THEOREM 14.1: If $\phi(y_1, \ldots, y_n)$ is a function of class A^1 in a region R and if for every $(y_1, \ldots, y_n) \in R$, $(y_1, \ldots, y_n, z = \phi(y_1, \ldots, y_n)) \in U$, then $z = \phi(y_1, \ldots, y_n)$ is a solution of (14.1) if and only if through every point of the surface $z = \phi(y_1, \ldots, y_n)$ there passes at least one characteristic curve (14.3) lying entirely on the surface.

Proof: Kamke[1], pp. 330-331. One can verify this directly.

For the homogeneous linear equation

$$(14.4) \quad \sum_{\nu=1}^{n} f^{\nu}(y_1,\ldots,y_n)\frac{\partial z}{\partial y_\nu} = 0$$

the characteristic equations become:

$$(14.5) \quad y_\nu' = f^{\nu}(y_1,\ldots,y_n) \quad (\nu=1,\ldots,n)$$

$$z' = 0$$

(usually when one speaks of the characteristic equations of (14.4) one omits $z' = 0$), and hence the characteristic curves are:

$$(14.6) \quad y_1 = \varphi_1(t),\ldots,y_n = \varphi_n(t), \; z = c$$

Again the term characteristic curve is often used for the projection of (14.6) on $z = 0$:

$$(14.7) \quad y_1 = \varphi_1(t),\ldots,y_n = \varphi_n(t).$$

The above theorem becomes:

THEOREM 14.2 If $f^{\nu}(y_1,\ldots,y_n)$ $(\nu=1,\ldots,n)$ are continuous in G and if $\varphi(y_1,\ldots,y_n)$ is a function of class A^1 in $R \leq G$, then $z = \varphi(y_1,\ldots,y_n)$ is a solution of (14.4) if and only if φ is constant along every characteristic curve, that is, if and only if $\varphi[\varphi_1(t),\ldots,\varphi_n(t)]$ = constant for every curve (14.7)

Proof: Kamke[1], p. 321-322. Easy to verify directly.

The solution of non-homogeneous equations (14.1) can be reduced to the solution of homogeneous linear equations (14.4) but in one higher dimension, by the following:

THEOREM 14.3 Let $f^{\nu}(y_1,\ldots,y_n,z)$ $(\nu=0,1,\ldots,n)$ be continuous in U

and let $\omega = \varphi(y,\ldots,y_n,z)$ be a solution in U of the homogeneous

equation:

$$(14.8) \quad \sum_{\nu=1}^{n} f^{\nu}(y_1,\ldots,y_n,z)\frac{\partial\omega}{\partial y_\nu} - f^0(y,\ldots,y_n,z) \cdot \frac{\partial\omega}{\partial z} = 0.$$

Suppose there exists a function $\varphi(y_1,\ldots,y_n)$ of class A^1 in R such that

for all $(y_1,\ldots,y_n)$ εR, $(y_1,\ldots,y_n,z = \varphi(y_1,\ldots,y_n))$ εU, and such that

$\psi(y_1,\ldots,y_n,\varphi(y_1,\ldots,y_n)) \equiv C$, some constant, in U, while $\psi_z(y_1,\ldots,y_n,$

$\varphi(y_1,\ldots,y_n)) \neq 0$ in any subregion $R_1 \leq R$. Then $z = \varphi(y_1,\ldots,y_n)$ is a

solution of the non-homogeneous equation (14.1) in R.

That is, if we obtain a solution of (14.8) and set $\psi(y_1,\ldots,y_n,$

$z) = c$ then if $\psi_z \neq 0$, this defines implicitly a function $z = \varphi(y_1,\ldots,y_n)$

which is a solution of (14.1).

Proof: See Kamke[1], p. 332. The result is found also in many other places.

The totality of all solutions of the homogeneous equation (14.4)

can be obtained from a system of (n-1) solutions called a *principal system*

of integrals in R of (14.4) these consist of any (n-1) functions.

$$(14.9) \quad \psi^1(y_1,\ldots,y_n), \psi^2(y_1,\ldots,y_n),\ldots, \psi^{n-1}(y_1,\ldots,y_n)$$

such that each $\psi^i(y_1,\ldots,y_n)$ is of class A^1 in R, and $z = \psi^i(y_1,\ldots,y_n)$

is a solution of (14.4) in R, and such that the matrix

$$\left\|\frac{\partial\psi^i}{\partial y_k}\right\| \quad (i=1,\ldots,n-1; \ k=1,\ldots,n)$$

has rank (n-1) in every subregion $R_1 \leq R$.

THEOREM 14.4 Let $f^{\nu}(y_1,\ldots,y_n)$ $(\nu=1,\ldots,n)$ be continuous in R and not

vanish simultaneously (identically) in any subregion $R_1 \leq R$. Then any n solutions of (14.4) of class A^1 are dependent on each other, and the totality of all solutions of (14.4) of class A^1 in R is just the totality of all functions of class A^1 which are dependent on the principal system of integrals (14.9).

Proof: See Kamke[1], pp. 322-324.

In a similar manner, the process of getting solutions of the equation (14.1) as described in theorem 14.3 yields all solutions by taking all possible ψ's and all possible C's, as is shown in the following:

THEOREM 14.5 Let $\overset{\nu}{f}(y_1,\ldots,y_n,z)$ $(\nu=0,1,\ldots,n)$ be continuous in U and suppose that for $\nu=1,\ldots,n$, $\overset{\nu}{f}(y_1,\ldots,y_n,z)$ do not vanish simultaneously at any point of U. Let the homogeneous equation (14.8) have a principal system of integrals in U: $\overset{1}{\psi}(y_1,\ldots,y_n,z),\ldots,\overset{n}{\psi}(y_1,\ldots,y_n,z)$.
If $\varphi(y_1,\ldots,y_n)$ is a function of class A^1 in R such that $z = \varphi(y_1,\ldots,y_n)$ is a solution of (14.1) in R, then for every bounded closed subregion $R_1 \leq R$, there exists a solution $\omega = r(y_1,\ldots,y_n,z)$ of the equation (14.8) which is not identically 0 in any subregion $U_1 \leq U$ and for which

$$r(y_1,\ldots,y_n,\varphi(y_1,\ldots,y_n)) \equiv 0 \text{ in } R_1.$$

Proof: Kamke[1], pp. 333-334.

For the homogeneous equation in two dimensions, which we can write:

$$(14.10) \qquad f^1(x,y)\,\frac{\partial z}{\partial x} + f^2(x,y)\,\frac{\partial z}{\partial y} = 0$$

the characteristic equations become:

$$(14.11) \quad x' = f^1(x,y), \; y' = f^2(x,y) \; (\text{and } z' = 0)$$

It will follow as a consequence of theorem 14.2 that if we have a solution $z = \varphi(x,y)$ of 14.10 any intersection of the surface $z = \varphi(x,y)$ with a plane $z = c$ is a characteristic curve and if a characteristic curve has one point in common with $z = \varphi(x,y)$, it lies on the surface. Furthermore, the system of principal integrals (14.9) reduces to a single function: $\psi(x,y)$ is a principal integral of (14.10) in R if it is of class A^1 in R, and $z = \psi(x,y)$ is a solution of (14.10) in R, and $\psi(x,y)$ is not constant in any subregion of R.

$$\left\| \frac{\partial\psi}{\partial x} \;\; \frac{\partial\psi}{\partial y} \right\| \text{ is of rank } 1$$

Theorem (14.4) means then that if f^1 and f^2 are not simultaneously 0 in any subregion $R_1 \leq R$, the totality of solutions of class A^1 of (14.10) is just the totality of functions dependent on $\psi(x,y)$ in R; if $X(x,y)$ is a solution of class A^1 of (14.10), then $\omega(X(x,y))$ is also a solution if $\omega(u)$ has domain including the values of $X(x,y)$, when $(x,y) \in R$.

The results of this section are not existence theorems, but as with Section 9 are useful adjuncts in getting solutions. It is evident also, that results on existence of principal integrals are fairly important, so we give those first. Such theorems are denoted by P; if equation is in normal form, we use PN.

15. Existence of Principal Integrals

The homogeneous linear equation in normal form is:

$$(15.1) \quad \frac{\partial z}{\partial x} + \sum_{\nu=1}^{n} f^{\nu}(x,y_1,\ldots,y_n) \frac{\partial z}{\partial y_\nu} = 0$$

Its characteristic equations are:

$$(15.2) \qquad y_\nu{}' = \overset{\nu}{f}(x, y_1, \ldots, y_n) \qquad (\nu=1,\ldots,n)$$

where as with general equations in normal form we set $x = t$ since $x' = 1$.

In Section 6 we found that there is a unique solution $y_\nu = \varphi_\nu(x)$ $(\nu=1,\ldots,n)$, passing through every point $(\xi, \eta_1, \ldots, \eta_n)$ of a region S. Indeed, if we write the solution for every such point:

$$(15.3) \qquad y_\nu = \overset{\nu}{\Phi}(x;\ \xi, \eta_1, \ldots, \eta_n) \quad (\nu=1,\ldots,n)$$

then we have the so-called *characteristic functions* of 15.2. By theorem 6.4 as functions of n + 2 variables they are of class A^1 in an appropriate region and the determinant $\left|\overset{\nu}{\Phi}_{\eta_k}\right| > 0$. Furthermore, they satisfy the differential equation:

$$\frac{\partial \overset{\nu}{\Phi}}{\partial \xi} + \sum_{\mu=1}^{n} \overset{\mu}{f}(\xi, \eta_1, \ldots, \eta_n) \cdot \frac{\partial \overset{\nu}{\Phi}}{\partial \eta_\mu} = 0 \quad (\nu=1,\ldots,n)$$

Hence we easily arrive at the following basic theorem:

THEOREM 15.1

(a) Let x_0 be a fixed point in L: $a<x<b$.

(b) Let $\overset{\nu}{f}(x, y_1, \ldots, y_n)$ $(\nu=1,\ldots,n)$ be bounded and continuous and let $\overset{\nu}{f}_{y_k}(x, y_1, \ldots, y_n)$ be continuous in:

$$R:\quad a<x<b;\ -\infty<y_1, y_2, \ldots, y_n < +\infty$$

Then there exist functions $\overset{\nu}{\psi}(x, y_1, \ldots, y_n)$ $(\nu=1,\ldots,n)$ such that:

(α) $\overset{\nu}{\psi}$ are of class A^1 in R

(δ) $\overset{1}{\psi}(x, y_1, \ldots, y_n), \ldots, \overset{n}{\psi}(x, y_1, \ldots, y_n)$ form a principal system of integrals of (15.1) in R. Furthermore, $\left|\frac{\partial \overset{\nu}{\psi}}{\partial y_k}\right| > 0$ in R. These principal integrals are given by:

$$(15.4) \qquad \overset{\nu}{\psi}(x, y, \ldots, y_n) \equiv \overset{\nu}{\Phi}(x^0;\ x, y, \ldots, y_n) \quad (\nu=1,\ldots,n)$$

where the Φ^{ν} are defined in (15.3).

Analogous to the lemma on Page 43 we have:

LEMMA: Let $f(x,y,\ldots,y_n)$ be continuous and bounded and f_{y_k} $(k=1,\ldots,n)$ be continuous in a region:

$$R_0: a<x<b, c_1<y_1<d_1,\ldots,c_n<y_n<d_n$$

Then for every δ with $0 < \delta < \min\limits_{\nu} \frac{d_\nu - c_\nu}{2}$, there is a function $\bar{f}(x,y_1,\ldots,y_n)$ defined in R which is continuous, bounded and has $\bar{f}_{y_k}$ continuous in R and coincides with f in R_0^1: $a<x<b$; $c_i+\delta<y_i<d_i-\delta$ $(i=1,\ldots,n)$.

If $|f_{y_k}| \leq A$ in R_0, then $|\bar{f}y_k| \leq A$ in R.

Proof: Kamke[1], p. 327.

Using this and theorem 15.1 we get:

THEOREM 15.2

Let $f^i(x,y_1,\ldots,y_n)$ be continuous and bounded and $f^i_{y_k}$ continuous in R_0, defined above. Let $a<x_0<b$, and let R_1 be: $a<x<b$; $\bar{c}_i<y_i<\bar{d}_i$ where $c_i<\bar{c}_i<\bar{d}_i<d_i$. Then there exists a principal system of integrals $\psi^{\nu}(x,y_1,\ldots,y_n)$ $(\nu=1,\ldots,n)$ of class A^1 in R_1 of (15.1) and with

$$\left|\frac{\partial\psi^{\nu}}{\partial y_k}\right| > 0 \text{ in } R_1.$$

(When n = 2, Rodebaugh[1] has extended this result to doubly and triply connected regions.)

When the equation is no longer in normal form, this result has been extended by Kamke[2] and Digel[2] for n = 2. Indeed, Kamke proved a result for non-homogeneous linear equations:

THEOREM 15.3: Let $f^1(x,y)$ and $f^2(x,y)$ be of class A^1 in R, and let $(f^1)^2 + (f^2)^2 > 0$ everywhere in R. Then in every connected subregion R_1 in the interior of R, in which f^1 and f^2 are bounded, there exists a principal integral $z = \psi(x,y)$ of (14.10), that is, a solution of (14.10) of class A^1 with $\psi_x^2 + \psi_y^2 > 0$ in R.

Proof: Kamke[2].

THEOREM 15.4: Let $f^1(x,y)$ and $f^2(x,y)$ be of class A^1 in a neighborhood N_δ of (0,0). Let the characteristic equations (14.11) have only closed integral curves of N. Then there exists a principal integral $z = \psi(x,y)$ of (14.10) in the deleted neighborhood N^1 (all of N except (0,0)).

Proof: Digel[2].

THEOREM 15.5: Let $f^i(x,y)$ $(i=0,1,2)$ be of class A^1 in R, and let $(f^1)^2 + (f^2)^2 > 0$ in R. Then in every subregion R_1 in the interior of R, where f^1 and f^2 Are bounded, there exists a function $\varphi(x,y)$ such that $z = \varphi(x,y)$ is a solution of:

$$f^1(x,y)\frac{\partial z}{\partial x} + f^2(x,y)\frac{\partial z}{\partial y} = f^0(x,y)$$

Proof: Kamke[2].

16. Solution of problems N and IN.

THEOREM 16.1: (a) Let x_0 be a given number in L: $a<x<b$ and let $g(y_1,\dots y_n)$ be a function of class A^1 in T: $c_i<y<d_i$ $(i=1,\dots n)$.

(b) Let $f^i(x,y_1,\ldots,y_n)$ be continuous, bounded, and let $f^i_{y_k}$ be continuous in $R_0 = L \times T$.

Then in every subregion $\bar{R}$: $a<x<b$, $\bar{c}_i<y_i<\bar{d}_i$, where $c_i<\bar{c}_i<\bar{d}_i<d_i$, there exists a function $\varphi(x,y_1,\ldots,y_n)$ such that;

(α_1) $\varphi(x,y_1,\ldots,y_n)$ is of class A^1 in $\bar{R}$

(β_1) $z = \varphi(x,y\ ,\ldots,y_n)$ is a solution of:

$$(14.1)\quad \frac{\partial z}{\partial x} + \sum_{\nu=1}^{n} f^{\nu}(x,y_1,\ldots,y_n)\frac{\partial z}{\partial y_\nu} = 0$$

(γ_0) For all y in $\bar{T}$: $\bar{c}_i<y_i<\bar{d}_i$ $\varphi(x_0,y_1,\ldots,y_n) \equiv g(y_1,\ldots,y_n)$

Proof: By lemma of previous section, extend f^{ν} to F^{ν} and g to G in the region R: $a<x<b$, $-\infty<y_1,\ldots,y_n<+\infty$. Apply theorem (16.1) and get principal integrals of $\frac{\partial z}{\partial x} + \Sigma F^{\nu}(x,y_1,\ldots,y_n)\frac{\partial z}{\partial y_\nu} = 0$. Calling them $\Phi(x_0;x,y_1,\ldots,y_n)$,

set $$\varphi(x,y_1,\ldots,y_n) \equiv G[\Phi(x_0;x,y_1,\ldots,y_n)]$$

Kamke[1], p. 328; pp. 314-317 for n=2.

For quasi-linear equations we get:

THEOREM 16.2

(a) Let x_0 be a given number and let $g(y_1,\ldots,y_n)$ be bounded and of class A^1 in T: $c_i<y_i<d_i$; let $|g_{y_k}| \geq \frac{1}{n} > 0$ in T.

(b) Let $f^i(x,y_1,\ldots,y_n,z)$ $(i=0,1,\ldots,n)$ be bounded and continuous with $f^i_{y_k}$ continuous in

$$U:\ |x-x_0| < a;\ c_i<y_i<d_i;\ -\infty<z<+\infty$$

Let $|f^i_{y_k}| \leq A$; $|f^i| \leq A$.

Then there exists a function $\varphi(x,y_1,\ldots,y_n)$ with the properties:

(α) $\varphi(x,y_1,\ldots,y_n)$ is of class A^1 in the region R:

$|x-x_0| < \alpha$; $\bar{c}_i<y_i<\bar{d}_i$ where $c_i<\bar{c}_i<\bar{d}_i<d_i$ and

$\alpha = \min\left(a, \frac{\log 2}{(n+1)A}\right)$

(β) $z = \varphi(x,y_1,\ldots,y_n)$ is a solution of (**14.1**) in R

(γ_0) $\varphi(x_0,y_1,\ldots,y_n) \equiv g(y_1,\ldots,y_n)$.

Proof: Kamke[1], pp. 335-340.

Theorem 16.1 has been extended by Kamke[4] to any homogeneous or non-homogeneous linear equation, where functions are of class A^m. We give results for n=2 as well as for the general case, because of their importance. The change of notation is due to the fact that the f^0 of 14.1 is now a sum of two terms. As in 16.1, the solution of problem N for linear equations does not have as its domain R a region depending on x_0; hence there is a uniformity with respect to x_0.

THEOREM 16.3:

(a_m) Let x_0 be any number in L_1: $a\leq x\leq b$, and let g(y) be a function of class A^m for all y. ($m\geq 1$)

(b_m) Let $f^i(x,y)$ (i=0,1) and $g^0(x,y)$ be of class A^m in R_1:

$a\leq x\leq b$; $-\infty<y<+\infty$

or more generally, let $f^i_y(x,y)$ and $g^0_y(x,y)$ be of class A^{m-1} in R_1. Let $f^i(x,y)$ be bounded. Then there exists a function $\varphi(x,y)$ such that

(α) $\varphi(x,y)$ is of class A^1 in R_1 .

(β) $z = \varphi(x,y)$ is a solution of the non-homogeneous equation:

(16.1) $\frac{\partial z}{\partial x} + f^1(x,y)\frac{\partial z}{\partial y} = f^0(x,y)z + g^0(x,y)$ in R_1

(γ) $\varphi(x_0,y) \equiv g(y)$ for all y.

Furthermore $\varphi(x,y)$ is of class A^m in R_1.

COROLLARY: If we let $z = \varphi(x,y;\xi)$ be the solution of the above problem for any ξ in L_1, (that is, ξ in place of x_0), then $\varphi(x,y;\xi)$ is of class A^m in the three-dimensional region $R_1 \times L_1$; $a \leq x \leq b$; $-\infty < y < +\infty$; $a \leq \xi \leq b$. It is given, parametrically by the formulas:

$$(16.2) \begin{cases} y = \Phi(x;\xi,\eta) \\ z = e^F \left[g(\eta) + \int_\xi^x g^0(\bar{x}, \Phi)\, e^{-F}\, d\bar{x}\right] \end{cases}$$

where $\Phi(x;\xi,\eta)$ is the solution of the equation:

(16.3) $y' = f^1(x,y)$

which passes through (ξ,η) and $F = F(x,\xi,\eta) = \int_\xi^x f^0(\bar{x}, \bar{\Phi})\, d\bar{x}$. If we solve the first of equations (16.2) for η, we get $\eta = \Phi(\xi;x,y)$. Substituting in the second expression, we would get $z = \varphi(x,y;\xi)$ by a single formula.

Proof: Kamke[4], pp. 276-277. The characteristic equations of 16.1 reduce to 16.3 and 16.4 $z' = f^0(x,y)\, z + g^0(x,y)$. The first equation, 16.3 has solution $y = \Phi(x;\xi,\eta)$ with all desired properties. (See theorem 15.1 Then substituting in 16.4 for y, we get an ordinary first order linear differential equation:

$$z' = f^0[x, \Phi(x;\xi,\eta)] \cdot z + g^0[x, \Phi(x;\xi,\eta)]$$

which is solved in the usual manner to give the second line of 16.2 Uniqueness and differentiability properties are then established.

The same result for n variables becomes:

THEOREM 16.4

(a_m) Let x_0 be any number in L_i: $a \leq x \leq b$ and let $g(y_1,\ldots,y_n)$ be a function of class A^m for all $y_1,\ldots,y_n$. $(m>1)$

(b_m) Let $f^i(x,y_1,\ldots,y_n)$ $(i=0,1,\ldots,m)$ and $g^o(x,y_1,\ldots,y_n)$ be of class A^m in R : $a\leq x\leq b$; $-\infty<y_1,\ldots,y_n<+\infty$; or, more generally, let $f^i_{y_k}(x,y_1,\ldots,y_n)$ and $g^o_{y_k}(x,y_1,\ldots,y_n)$ be of class A^{m-1} in R_1. $(i=0,1,\ldots,n)$, $k=1,\ldots,n)$. Let $f^i(x,y_1,\ldots,y_n)$ be bounded $(i=1,\ldots,n)$. Then there exists a function $\varphi(x,y_1,\ldots,y_n)$ such that:

(α_1) $\varphi(x,y_1,\ldots,y_n)$ is of class A^1 in R_1.

(β_1) $z = \varphi(x,y_1,\ldots,y_n)$ is a solution of the non-homogeneous equation: (16.5)

$$\frac{\partial z}{\partial x} + \sum_{k=1}^{n} f^i(x,y_1,\ldots,y_n)\frac{\partial z}{\partial y_k} = f^o(x,y_1,\ldots,y_n)z + g^o(x,y_1,\ldots,y_n) \text{ in } R_1.$$

(γ_0) $\varphi(x_0,y_1,\ldots,y_n) \equiv g(y_1,\ldots,y_n)$ for all y.

Furthermore, $\varphi(x,y_1,\ldots,y_n)$ is of class A^m in R_1. (Kamke)[4]

COROLLARY: If we let $z = \varphi(x,y_1,\ldots,y_n;\xi)$ be the solution of the above problem for any ξ in L_1, then φ is a function of class A^m in the n+2 dimensional region $R_1 \times L_1$. It is given parametrically by the formulas:

$$(16.6)\quad \begin{cases} y_i = \Phi^i(x;\xi,\eta_1,\ldots,\eta_n) & (i=1,\ldots,n) \\ z = e^F[g(\eta_1,\ldots,\eta_n) + \int_\xi^x g^o(\bar{x}, \bar{\Phi}^1,\ldots, \bar{\Phi}^n)\, e^{-\bar{F}} d\bar{x}] \end{cases}$$

where $\Phi^i(x;\xi,\eta_1,\ldots,\eta_n)$ are the solutions of

$$(15.2)\quad y_i' = f^i(x,y_1,\ldots,y_n) \quad (i=1,\ldots,n)$$

which pass through a given $(\xi,\eta_1,\ldots,\eta_n)$.

Since $J = \left|\frac{\partial\Phi^i}{\partial y_k}\right| \neq 0$

the first n equations can be solved for η_i in terms of y_i and ξ and substituted in the last one to get $z = \varphi(x, y_1, \ldots y_n; \xi)$. Theorems 16.2 and 16.3 can be applied to finite regions by using the lemma of Section 14. For solutions of (16.5) depending on a parameter, see Kamke[4].

The following result, given for problem IN, is sometimes useful:

THEOREM 16.5 (b) Let $f(x,y)$, $g(x,y)$, $f_y(x,y)$, $g_y(x,y)$ be continuous in

$$(B_0) \quad 0 \le x \le a, \quad |y| + K\,x \le b$$

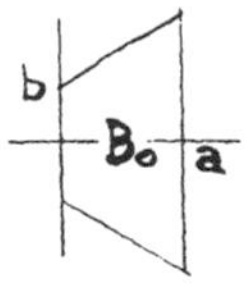

and suppose: $|f| \le K$, $|f_y| \le L$

$$|g| \le \frac{Mx^k}{k!}, \quad |g_y| \le \frac{Nx^j}{j!}$$

where a, b, K, L, M, N are positive constants and $k \ge 0$, $j \ge 0$. Then there exists one and only one function $\varphi(x,y)$ such that:

(α) $\varphi(x,y)$ is of class A^1 in B_0 and

$$|\varphi| \le \frac{Mx^{k+1}}{(k+1)!}, \quad |\varphi_y| \le \frac{Ne^{La}x^{j+1}}{(j+1)!} \quad \text{in } B_0.$$

(β_1) $z = \varphi(x,y)$ is a solution of the equation:

$$\frac{\partial z}{\partial x} = f(x,y)\frac{\partial z}{\partial y} + g(x,y) \text{ in } B_0.$$

(γ_0) $\varphi(0,y) \equiv 0$ for $|y| \le b$.

Proof: Perron[2]. As in theorem 16.2.

$$\varphi(x,y) = \int_0^x g[t, \varphi(t, \psi(x,y))]\, dt$$

where $y = \varphi(x,c)$ is the solution of $\frac{dy}{dx} + f(x,y) = 0$ for which $\varphi(0,c) = c$.

Solving $y = \varphi(x,c)$, we get $c = \psi(x,y)$.

17. Solution of quasi-linear equations, not in normal form.

As in sections 12 and 13, when a quasi-linear equation is not given in normal form, one can reduce it to normal form by applying the transformations of section 8. This is simple in case of problems G, I, $\overline{C}$. In the case of problem C, the projection of the initial manifold Γ(13.5) is

$$\Gamma_0: \quad x = x^{\circ}(\mu_1, \ldots \mu_n); \; y = y_i^{\circ}(\mu_1, \ldots, \mu_n) \qquad [i=1, \ldots, n]$$

where the functional matrix J is of rank n, and all functions are of class A^1 in some set M. Furthermore, (13.6) reduces to

$$\Delta = \begin{vmatrix} f_0(P) & f_1(P) & \ldots & f_n(P) \\ \frac{\partial x_0}{\partial \mu_1} & \frac{\partial y_1}{\partial \mu_1} & \ldots & \frac{\partial y_n}{\partial \mu_1} \\ \cdot & \cdot & \ldots & \cdot \\ \frac{\partial x_0}{\partial \mu_n} & \frac{\partial y_1}{\partial \mu_n} & \ldots & \frac{\partial y_n}{\partial \mu_n} \end{vmatrix} \qquad (P \varepsilon \Gamma)$$

THEOREM 17.1: Let $x^{\circ}(\mu_1, \ldots, \mu_n)$, $y_i^{\circ}(\mu_1, \ldots, \mu_n)$, $\ldots, z^{\circ}(\mu_1, \ldots, \mu_n)$ be functions of class A^1 in a set M, and let $J = \left\| \frac{\partial x^{\circ}}{\partial \mu_k} \; \frac{\partial y_i^{\circ}}{\partial \mu_k} \; \frac{\partial z^{\circ}}{\partial \mu_k} \right\|$ be of rank n in M. Let $f_j(x, y_1, \ldots, y_n, z)$ be of class A° in $G = G_0 \times T$, where G_0 is a simply-connected region in $(x, y_1, \ldots, y_n)$ -space and T is an open interval: $|z| < L$. For all $(\mu_1, \ldots \mu_n)$ in M, let the points of Γ lie in G. Let $\Delta \neq 0$ for all $\mu \varepsilon M$.

Then there exists a function $\varphi(x, y_1, \ldots y_n)$ such that

(α) $\varphi(x, y_1, \ldots y_n)$ is defined and of class A^1, in R, a subregion of G_0.

(β) $z = \varphi(x, y_1, \ldots y_n)$ is a solution in R of

$$(17.2) \quad f_0(x, y_1, \ldots, y_n, z) \frac{\partial z}{\partial x} + \sum_{j=1}^{n} f_j(x, y_1, \ldots y_n, z) \frac{\partial z}{\partial y_j} = f_{n+1}(x, y_1, \ldots, y_r$$

(γ) Γ_0 lies in R and $\varphi[x^{\circ}(\mu_1, \ldots \mu_n), y_1^{\circ}(\mu_1, \ldots \mu_n), \ldots, y_n^{\circ}(\mu_1, \ldots \mu_n)] \equiv z^{\circ}(\mu_1, \ldots$

(Statement of this theorem in Courant-Hilbert, II, p. 59 - 2 dimensions on p. 54.)

Again if $\Delta = 0$ along Γ, then if the problem has a solution, Γ is a characteristic manifold, and there are an infinite number of solutions. This theorem does not give the region of existence of the solution, as a reduction to normal form does. Because of its importance, we give the result for n=1:

COROLLARY: Let C_0 $\begin{cases} x=x_0(\mu) \\ y=y_0(\mu) \end{cases}$ be a curve for which $[x_0^{(1)}(\mu)]^2 + [y_0^{(1)}(\mu)]^2 \neq 0$ for all $\mu \varepsilon M$, and where $x_0(\mu)$ and $y_0(\mu)$ are of class A^1 in M. (C has no double points).

Let $z_0(\mu)$ be a function of class A^1 in M.

Let $f_1(x,y,z)$, $f_2(x,y,z)$, $f_3(x,y,z)$ be of class A^0, for $(x,y) \varepsilon G_0$, where G_0 is a simply connected region and $z \varepsilon T$: $|z| < Z$.

Set $G = G_0 \times T$ and for all $\mu \varepsilon M$, assume $[x_0(\mu), y_0(\mu), z_0(\mu)] \varepsilon G$.

Let $\Delta \equiv f_1[x_0(\mu)\Gamma\ y_0(\mu)] \cdot y_0'(\mu) - f_2[x_0(\mu),\ y_0(\mu)] \cdot z_0'(\mu) \neq 0$ for all $\mu \varepsilon M$.

Then there exists a subregion $R \leq G_0$ which contains C_0, and a unique function $\varphi(x,y)$ defined for all $(x,y) \varepsilon R$ such that

(α) $\varphi(x,y)$ is of class A^1 in R

(β) $z = \varphi(x,y)$ is a solution of

$$f_1(x,y,z)\frac{\partial z}{\partial x} + f_2(x,y,z)\frac{\partial z}{\partial y} = f_3(x,y,z) \text{ in } R$$

(γ) $\varphi(x_0(\mu),\ y_0(\mu)) \equiv z_0(\mu)$ on C_0.

E. Systems of First Order Equations

18. In this section, the system of r equations in r unknown functions and n+1 independent variables:

$$(18.1)\quad F_j(x, y_1, \ldots, y_n, z_1, \ldots, z_r, \frac{\partial z_1}{\partial x}, \ldots, \frac{\partial z_r}{\partial x}, \frac{\partial z_1}{\partial y_1}, \ldots, \frac{\partial z_r}{\partial y_n}) = 0 \quad (j=1, \ldots,$$

is considered. If the functions F_j are of class A^0 at every point $P: (x, y_1, \ldots y_n, z_1, \ldots z_r, p_1, \ldots p_r, q_{11}, \ldots, q_{rn})$ of a region U, if $\frac{\partial F_j}{\partial p_i}$ are of class A^0 in U and the determinant

$$(18.2)\quad J \equiv \left| \frac{\partial F_j}{\partial p_i} \right| \neq 0$$

and if for some P^0 in U, $F_j(P^0) = 0$, then by the implicit function theorem,

$$F_j(P) = 0$$

can be solved for $p_1, \ldots, p_r$ to give:

$$p_j = f_j(x, y_1, \ldots y_n, z_1, \ldots z_r, q_{11}, \ldots q_{rn})$$

where $\overset{\circ}{p}_j = f_j(x^0, \overset{\circ}{y}_i, \overset{\circ}{z}_j, \overset{\circ}{q}_{ik})$. (The purpose of the point P^0 is to obtain a unique solution.) Hence the solution of equation (18.1) is equivalent to the solution of the normal form:

$$(18.3)\quad \frac{\partial z_j}{\partial x} = f^j(x, y_1, \ldots y_n, z_1, \ldots z_r, \frac{\partial z_1}{\partial y_1}, \ldots, \frac{\partial z_r}{\partial y_n}) \qquad (j=1, \ldots r)$$

If (18.2) is not true throughout U, it may be that for some k', the determinant

$$(18.4)\quad \left| \frac{\partial F_j}{\partial q_{ik}} \right| \neq 0$$

whence one could interchange the labels on the variables x and $y_{k'}$ and again reduce (18.1) to a normal form. If neither (18.2) nor (18.4) holds throughout U, one may still effect a reduction to normal form in some subregion of U; indeed, it sometimes happens that U can be divided up into a finite number of overlapping subregions in each of which either (18.2) holds or else, for some k', (18.4) is true. More common, however, is the condition that either (18.2) or (18.4) hold at every point P^{o} of U, and hence that in a neighborhood of P^{o}, the equation (18.1) can be reduced to some normal form.

Problem C would be to determine a solution

$$z_j = \varphi_j(x, y_1, \ldots, y_n)$$

of the system (18.1), such that for a given set of functions

$$(18.5) \quad x = x^{o}(\mu_1, \ldots, \mu_n); \; y_i = \overset{o}{y}_i(\mu_1, \ldots, \mu_n); \; z_j = \overset{o}{z}_j(\mu_1, \ldots, \mu_n),$$

of class A^1 in a region M, where $\left\| \frac{\partial x^{o}}{\partial \mu_k} \frac{\partial y_1}{\partial \mu_k} \cdots \frac{\partial y_n}{\partial \mu_k} \right\|$ is of rank n in M,

$$\varphi_j(x^{o}(\mu), \overset{o}{y}_1(\mu), \ldots, \overset{o}{y}_n(\mu)) \equiv z_j(x^{o}(\mu), \ldots, \overset{o}{y}_n(\mu)) \text{ in M.}$$

Problem $\bar{C}$ would replace (18.5) by an initial manifold in which the y_i's are parameters:

$$x = h^{o}(y_1, \ldots, y_n); \; z_j = \overset{o}{g}_j(y_1, \ldots, y_n)$$

Problem G would be obtained from this by making the surface $x = h(y)$ the plane $x = x_0$. Problem I would be obtained from $\bar{C}$ by setting $h \equiv 0$ and each $g_j \equiv 0$. Problem N would be problem G for the normal form (18.3) and problem IN, problem I for the same equation. As in the case of a single equation (see section 8), one can show that the solutions of problems $\bar{C}$, G, I are equivalent, that a solution of N is equivalent to a solution of G provided (18.2) holds, and that a solution of $\bar{C}$ leads to a solution of C provided that for (18.5), $\frac{\partial y_i}{\partial \mu_k} \neq 0$. When the conditions do not hold throughout a region, it may still be possible to make the reduction of problem C or C to N in the neighborhood of every point of a region and so use the existence theorems below to obtain local solutions. (See Goursat[1], Germay[1])

THEOREM 18.1: (a_∞) Let $(x^{o}, \overset{o}{y}_1, \ldots, \overset{o}{y}_n)$ be a given point and $g^i(y_1, \ldots, y_n)$ $(i = 1, 2, \ldots, r)$ be functions of class A^∞ in a neighborhood T_δ of this point. Set $z_i^{o} = g^i(y_1^{o}, \ldots, y_n^{o})$ and $\overset{o}{q}_{ik} = g^i_{y_k}(y_1^{o}, \ldots, y_n^{o})$ $(i=1, \ldots, r;\; k=1, \ldots, n)$

(b_∞) Let $f^i(x, y_1, \ldots, y_n, z_1, \ldots, z_r, q_{11}, \ldots, q_{ik}, \ldots, q_{rn})$ $(i=1, \ldots, r)$

be functions of class A^{∞} in a neighborhood S_{δ} of $(x^{o}, y_1, \ldots, y_n)$. Then there exists a unique set of functions $\varphi^{i}(x, y_1, \ldots, y_n)$ $(i=1,\ldots,r)$ such that

(α_{∞}) $\varphi^{i}(x, y_1, \ldots, y_n)$ is of class A^{∞} in a neighborhood R_{δ} of $(x^{o}, y_1^{o}, \ldots, y_n^{o})$ $(i=1,\ldots,r)$

(β_0) $z_i = \varphi^{i}(x, y_1, \ldots, y_n)$ is a solution of:

$$(18.3)\quad \frac{\partial z_i}{\partial x} = f(x, y_1, \ldots, y_n, z, \ldots, z_r, \frac{\partial z_1}{\partial y_1}, \ldots, \frac{\partial z_i}{\partial y_k}, \ldots, \frac{\partial z_r}{\partial y_n}).$$

$$(i=1,\ldots,r)$$

(γ_0) $\varphi^{i}(x^{o}, y_1, \ldots, y_n) = g^{i}(y_1, \ldots, y_n)$ for all $y \varepsilon T_{\delta_1}$. $(i=1,\ldots,r)$

Proof: Goursat[1], pp. 2-6. The method is exactly the same as that used for theorem 10.1.

Perron[1] has given a result for this system analogous to the corollary to theorem 10.1.

THEOREM 18.2: Let (a_{∞}) and (b_{∞}) be given as above. Then there exist r uniformly convergent sequences of functions

$$\varphi^{ij}(x, y_1, \ldots, y_n) \quad (j=0,1,2,\ldots;\ i=1,\ldots,r)$$

of class A^{∞} in a neighborhood of $(x_0, y_1^{o}, \ldots, y_n^{o})$ such that

$$\lim_{j\to\infty} \varphi^{ij}(x, y_1, \ldots, y_n) = \varphi^{i}(x, y_1, \ldots, y_n) \quad (i=1,\ldots,r)$$

satisfy (α_{∞}), (β_0), (γ_0), and hence are the unique functions of theorem 10.1.

Proof: Germay[1], pp. 31ff. The method of proof is the same as that for theorem 10.2.

The quasi-linear equation system, where $F^j_{ik} = F^j_{ik}(x, y_1, \ldots y_n, z_1, \ldots z_r)$:

$$(18.6) \quad \sum_{i=1}^{r} F^j_{io} \frac{\partial z_i}{\partial x} + \sum_{k=1}^{n} \sum_{i=1}^{r} F^j_{ik} \frac{\partial z_i}{\partial y_k} = F^j_o$$

can be reduced to normal form:

$$(18.7) \quad \frac{\partial z_j}{\partial x} + \sum_{k=1}^{n} \sum_{i=1}^{r} f^j_{ik} \frac{\partial z_i}{\partial y_k} = f^j_o$$

provided either (18.2) or (18.4) is satisfied, that is, provided

$$\left| F^j_{io} \right| \neq 0 \quad \text{or} \quad \left| F^j_{ik'} \right| \neq 0$$

in a region U, or in some subregion of U.

If one can establish the existence of a solution of problem N for the quasi-linear system (18.7) one can solve the corresponding problem N for the non-linear system (18.4) by finding the solution of the quasi-linear system in unknowns z_j, q_{ik}:

$$\frac{\partial z_j}{\partial x} = f^j(x, y_1, \ldots y_n, z_1, \ldots z_r, q_{11}, \ldots q_{rn}); \quad \frac{\partial q_{ik}}{\partial x} = f^i_{x_k} + \sum_{h=1}^{r} f_{z_h} \frac{\partial z_h}{\partial y_k}$$

$$+ \sum_{h=1}^{r} \sum_{m=1}^{n} f^i_{q_{km}} \cdot \frac{\partial q_{hm}}{\partial y_k}$$

with auxiliary conditions: $z_j(x^o, y), \cdot\, y_n) = g^j(y_1, \ldots y_n) \cdot q_{jk}(x^o, y_1, \ldots y_n)$

$$= g^j_{y_k}(y_1, \ldots y_n)$$

and then showing that for the solution: $z_j = \varphi_j(x,y)$, $q_{ik} = \varphi_{ik}(x,y), \varphi_{ik} = \frac{\partial \varphi_i}{\partial y_k}$.

Then $z_j = \varphi_j(x,y)$ is the solution to the given problem. Indeed, this was Kowalewski's original method in establishing theorem 18.1.

Nagumo[1] used this method to establish the existence of a solution of problem N for (18.3), where the functions f^i are of class A^o in x and of class A^∞ in the other variables; he takes x real and the others complex, In order to prove the existence of a solution for (18.7) he makes use of Schauder's fixed-point theorem.

19. For 2 independent variables, equations (18.1) become:

$$(19.1)\quad F^i(x,y,z_1,\ldots,z_r,\frac{\partial z_1}{\partial x},\ldots,\frac{\partial z_r}{\partial x},\frac{\partial z_1}{\partial y},\ldots,\frac{\partial z_r}{\partial y}) = 0 \qquad (i=1,\ldots,r)$$

When all $F^i(x,y,z_1,\ldots,z_r,p_1,\ldots,p_r,q_1,\ldots,q_r)$ are of class A^1 in U and either $\left|F^i_{p_k}\right| \neq 0$ or $\left|F^i_{q_k}\right| \neq 0$ throughout U, the system can be reduced to the normal form:

$$(19.2)\quad \frac{\partial z^i}{\partial x} = f^i(x,y,z_1,\ldots,z_r,\frac{\partial z_1}{\partial y},\ldots,\frac{\partial z_r}{\partial y}) \qquad (i=1,\ldots,r)\quad .$$

When $F^i_{p_k} \neq 0$ or $F^i_{q_k} \neq 0$ at every point of U, a reduction to normal form can be made in the neighborhood of each point.

Associated with (19.1) is an equation called the characteristic equation:

$$\left|F^i_{p_k}\alpha_1 + F^i_{q_k}\right| = 0$$

which for the normal form (19.2) becomes:

$$(19.3)\quad \begin{vmatrix} f_{11}-\alpha & f_{12} & \ldots & f_{1n} \\ f_{21} & f_{22}-\alpha & \ldots & f_{2n} \\ \cdot & \cdot & \cdot & \cdot \\ f_{n1} & f_{n2} & \ldots & f_{nn}-\alpha \end{vmatrix} = 0 \qquad (f_{ij} = f^i_{g_j})$$

For a fixed point P in U, this will have n roots: $\alpha=\alpha_1,\alpha_2,\ldots,\alpha_n$. If they are all distinct and real, (9.1) is said to be hyperbolic at P; if not, the terminology as to whether the system is called elliptic or parabolic at P is not always uniform. If the system is hyperbolic at every point of a region U, it is called hyperbolic in U.

For the quasi-linear system;

$$(19.4)\quad \sum_{\kappa=1}^{r} a_{i\kappa}(x,y,z_1,\ldots z_r)\frac{\partial z_\kappa}{\partial x} + \sum_{\kappa=1}^{r} b_{i\kappa}(x,y,z_1,\ldots,z_r)\frac{\partial z_\kappa}{\partial y}$$
$$= c_i(x,y,z_1,\ldots,z_r)$$

which in normal form becomes:

$$(19.5)\quad \frac{\partial z_\kappa}{\partial \kappa} + \sum_{\kappa=1}^{r} b_{i\kappa}(x,y,z_1,\ldots,z_r)\frac{\partial z_\kappa}{\partial y} = c_i(x,y,z_1,\ldots,z_r).$$

the characteristic equation depends only upon: $(x, y, z_1, \ldots z_r)$, so one can speak of the system being hyperbolic, or not, at a point Q: $(x, y, z_1, \ldots z_r)$ or in a region S of the $(x, y, z_1, \ldots, z_r)$-space. Finally, for the semi-linear equation:

$$(19.6)\quad \sum_{\kappa=1}^{r} a_{i\kappa}(x,y)\,\frac{\partial z_\kappa}{\partial \kappa} + \sum_{\kappa=1}^{r} b_{i\kappa}(x,y)\,\frac{\partial z_\kappa}{\partial y} = c_i(x, y, z_1, \ldots, z_r)$$

which in normal form becomes

$$(19.7)\quad a_{i\kappa}\,\frac{\partial z_\kappa}{\partial x} + \sum_{\kappa=1}^{r} b_{i\kappa}(x,y)\,\frac{\partial z_\kappa}{\partial y} = c_i(x, y, z_1, \ldots, z_\kappa)$$

the characteristic equation becomes:

$$|a_{i\kappa}(x,y)\cdot\alpha + b_{i\kappa}(x,y)| = 0$$

which depends only upon (x,y) so that one can consider the system (19.6) as being hyperbolic or not at a point of a region R of the (x,y) plane. Indeed, in this case, any curve: $x = x_0(\mu)$, $y = y_0(\mu)$ through the point (x^0, y^0) such that $\frac{dy}{dx} = \alpha_i$ at the point is said to have a characteristic direction at (x^0, y^0). When (19.6) is in the special form which has only one unknown in each equation:

$$(19.8)\quad a_\kappa(x,y)\,\frac{\partial z_\kappa}{\partial x} + b_\kappa(x,y)\,\frac{\partial z_\kappa}{\partial y} = c_\kappa(x,y) \qquad (\kappa = 1, \ldots, r)$$

the characteristic determinant is in diagonal form, so the characteristic equation becomes: $\prod_{\kappa=1}^{r}(a_\kappa\alpha + b_\kappa) = 0$ and, when $a_\kappa \neq 0$ the roots of this equation are just $-\alpha_\kappa = -\dfrac{b_\kappa}{a_\kappa}$.

Holmgren[1] showed the existence of a unique solution of the problem of Cauchy for the homogeneous linear system, obtained from (19.7) when $c_i = \sum_{\kappa=1}^{r} c_{i\kappa}(x,y)\cdot z_\kappa$ in case the initial curve C_0: $x = x_0(\mu), y = y_0(\mu)$ was not tangent to a characteristic direction at any point and the equation was hyperbolic. KFor the case $r = 2$, Carleman[1] obtained some results for uniqueness of solution in non-hyperbolic cases.) Perron extended the work of Holmgren to obtain the existence of a solution of problem IN, in the large, for (19.6), by first considering a system of the form (19.8); the results are given in the next two theorems.

THEOREM 19.1

(b) Let $\overset{\nu}{F}(x,y)$ and $\overset{\nu}{G}(x,y,z_1,\ldots,z_n)$ $(\nu=1,\ldots,n)$ be continuous in B_0: $0\leq x\leq a$, $|y| + Kx<b$

and B_1: $0\leq x\leq a$, $|y| + Kx<b$, $|z_\nu| <c$ $(\nu=1,\ldots,n)$ respectively, where a, b, c, K are positive constants and $|\overset{\nu}{F}(x,y)| \leq K$. Let $|\overset{\nu}{G}| \leq M$, for some constant M. Let $\overset{\nu}{F}_y$ exist and be continuous in B_0 and $\overset{\nu}{G}_y$, $\overset{\nu}{G}_{z_\lambda}$, $\overset{\nu}{G}_{y,z_\lambda}$ $\overset{\nu}{G}_{z_\lambda,z_\mu}$ be continuous in B_1 $(\lambda,\mu,\nu=1,\ldots,n)$. Then there is exactly one set of functions $\varphi_\nu(x,y)$ $(\nu=1,\ldots,n)$ such that:

(α) $\varphi_\nu(x,y)$ is of class A^1 in

B_2: $0\leq x\leq\alpha$, $|y| + Kx<b$ where $\alpha = \min(a, \frac{c}{M})$ and

$|\varphi_\nu(x,y)| \leq c$

(β) $z_\nu = \varphi_\nu(x,y)$ is a solution of the differential equation system:

$$(19.9)\quad \frac{\partial z_\nu}{\partial x} = F_\nu(x,y)\,\frac{\partial z_\nu}{\partial y} + G_\nu(x,y,z_1,\ldots,z_n) \quad (\nu=1,2,\ldots,n)$$

(γ) $\varphi_\nu(0,y) \equiv 0.$

Proof: Perron [2], pp. 557-562.

Set $\varphi_{\nu,0} = 0$ $(\nu=1,\ldots,n)$. Then define $\varphi_{\nu,m+1}(x,y)$, by a recursion relation, namely, that it satisfy

$$\frac{\partial\varphi_{\nu,m+1}}{\partial x} = \overset{\nu}{F}(x,y)\,\frac{\partial\varphi_{\nu,m+1}}{\partial y} + \overset{\nu}{G}(x,y,\varphi_{1,m},(x,y),\ldots,\varphi_{n,m},(x,y))$$

$$(\nu=1,\ldots,n)$$

Because of the result on page 61 this is possible, and indeed, it turns out that

$$|\varphi_{\nu,m+1} - \varphi_{\nu,m}| = |v_{\nu,m+1}| \leq \frac{C_1^m x^m}{m!} \quad \text{in } B_2.$$

Hence,

$$\varphi(x,y) = \lim_{m\to\infty} \varphi_{\nu,m}(x,y) = \sum_{m=1}^{\infty} v_{\nu,m}(x,y)$$ exists, since the series converges absolutely and indeed uniformly. This will be the desired system.

This result can be generalized: (Perron[2]).

THEOREM 19.2

(a) Let $f^{\nu\mu}(x,y)$ be of class A^2 (except that $\frac{\partial^2 f^{\nu\mu}}{\partial x^2}$ need not exist) in the region $(\nu=1,\ldots,n)$

(B_2) $0\leq x\leq a$, $|y| \leq b$ and let $g^{\nu}(x,y,z_1,\ldots,z_n)$ be of class A^2 (except that $\frac{\partial^2 g^2}{\partial x \partial y}$, $\frac{\partial^2 g^2}{\partial x \partial z_\lambda}$ need not exist) in the region

(B_3) $0\leq x\leq a$, $|y| \leq b$, $|z_\nu| \leq c$ where a, b, c are positive constants.

(c) Let the characteristic equations:

$$\begin{vmatrix} f^{11}-\alpha & f^{12} & \ldots & f^{1n} \\ f^{21} & f^{22}-\alpha & \ldots & f^{2n} \\ \ldots & \ldots & \ldots & \ldots \\ f^{n1} & f^{n2} & \ldots & f^{nn}-\alpha \end{vmatrix}$$

have real distinct roots and let these roots $F^{\nu}(x,y)$ $(\nu=1,\ldots,n)$ have

$$|F^{\nu}(x,y)| \leq K.$$

Then there exist functions $\varphi_\nu(x,y)$ such that;

(α) $\varphi_\nu(x,y)$ is of class A^1 in

(B_4) $0\leq x\leq a^1$, $|y| + Kx\leq b$ where $0<a^1<a$ and $|\varphi_\nu| \leq c$

(β) $z_\nu = \varphi_\nu(x,y)$ $(\nu=1,\ldots,n)$ satisfies the system:

$$\frac{\partial z_\nu}{\partial x} = \sum_{\mu=1}^{n} f_{\nu\mu}(x,y)\frac{\partial z_\mu}{\partial y} + g_\nu(x,y,z_1,\ldots,z_n) \quad (\nu=1,2,\ldots,n)$$

(γ) $\varphi_\nu(0,y) \equiv 0.$

The solution is unique.

Proof: One can find a matrix $||P_{ij}||_{n\times n}$ such that

$$\left\|P_{ij}\right\| \cdot \left\|f_{ij}\right\| = \left\|\begin{matrix}\alpha_1 & & 0\\ & \alpha_2 & \\ & & \ddots \\ 0 & & \alpha_n\end{matrix}\right\| \cdot \left\|P_{ij}\right\|$$

where $\alpha_1,\ldots,\alpha_n$ are the roots of the characteristic equation and such that P_{ij} and its derivatives are continuous in R. Hence $\sum_{\nu=1}^{n} P_{\lambda\nu} f_{\nu\mu} = \alpha_\lambda \cdot P_{\lambda\mu}$.

Under the transformation $\mu_\lambda = \Sigma P_{\lambda\mu} z_\mu$ which has inverse $z_\mu = \Sigma Q_{\nu\lambda}\mu_\lambda$, a solution of (19.7) would transform into a solution of

$$\frac{\partial\mu_\lambda}{\partial x} = \alpha_\lambda \frac{\partial\mu_\lambda}{\partial y} + \sum_{\nu=1}^{n}\left(P_{\lambda\nu}g_\nu + \frac{\partial P_{\lambda\nu}}{\partial x}\mu_\lambda - \alpha_\lambda\frac{\partial P_{\lambda\nu}}{\partial y}\mu_\nu\right)$$

which is an equation of the type considered in the previous theorem.

In the non-normal case, we first give an existence theorem which does not deal with problem C but is analogous to the theorems for principal integrals given in section 15. The equation (19.9) is the case r = 2 of (19.8), so that its characteristic equation is $(f_1^1\alpha+f_1^2)(f_2^1\alpha+f_2^2) = 0$ which has two distinct roots provided $f_1^1 f_2^2 - f_2^1 f_1^2 \neq 0$; thus the hypothesis of the theorem insures that the equation is hyperbolic throughout R.

THEOREM 19.3: Let $f_\nu^{\,i}(x,y)$ $(i=0,1,2;\ \nu=1,2)$ be of class $A^m (m\geq 1)$ in R, a simply-connected region. Let $f_1^{\,1} f_2^{\,2} - f_2^{\,1} f_1^{\,2} > 0$ in R. Then in every bounded region R_1 in the interior of R, there exist functions $\varphi_\nu(x,y)$

$(\nu=1,2)$ which are of class A^k and such that $z_\nu = \varphi_\nu(x,y)$ are solutions in R_1 of:

$$(19.10) \qquad f_\nu^{\,1}(x,y)\,\frac{\partial z_\nu}{\partial x} + f_\nu^{\,2}(x,y)\,\frac{\partial z_\nu}{\partial y} = f_\nu^{\,0}(x,y) \qquad (\nu=1,2)$$

COROLLARY: Under the hypotheses of the theorem, in every R_1, there exist functions $U(x,y)$ and $V(x,y)$ such that $U_xV_y - V_xU_y > 0$ in R_1, U and V are of class A^m in R_1, $z_1 = U(x,y)$ and $z_2 = V(x,y)$ are solutions of:

$$f_\nu^{\,1}(x,y)\,\frac{\partial z}{\partial x} + f_\nu^{\,2}(x,y)\,\frac{\partial z}{\partial y} = 0 \qquad (\nu=1,2)$$

and the region R_1 is mapped in a 1-1 fashion on a region of the U-V plane by $z_1 = U(x,y)$, $z_2 = V(x,y)$.

Proof: Kamke[3].

The next theorem applied to a special case of (19.4). Some details of the proof are given because the method has been used in many situations.

THEOREM 19.4: (a) Let $a(z_1,z_2)$, $b(z_1,z_2)$, $c(z_1,z_2)$, $d(z_1,z_2)$, $e(z_1,z_2)$, $f(z_1,z_2)$ be functions of class A^1 in the interior of a region S.

(b) Let the line segment L in the (x-y) plane be defined parametrically in terms of arc length as follows:

$$\begin{cases} x = -\dfrac{\lambda}{\sqrt{2}} & \text{for all } \lambda \varepsilon M \text{ where } M\text{: } \lambda_1 \le \lambda \le \lambda_2 \text{ And } \lambda_1 < 0 < \lambda_2 \\ y = \dfrac{\lambda}{\sqrt{2}} & \end{cases}$$

[Hence $x+y = 0$, $x \leq \frac{-\lambda_1}{\sqrt{2}} = x_0$, $y \leq \frac{\lambda_2}{\sqrt{2}} = y_0$, $x_0 + y_0 = \frac{\lambda_1 - \lambda_2}{\sqrt{2}} > 0$]

(c) Let $z_1^{\circ}(\lambda)$ and $z_2^{\circ}(\lambda)$ be functions of class A^1 in M, such that $(z_1^{\circ}(\lambda), z_2^{\circ}(\lambda)) \in S$ and such that

$$\Delta = \begin{vmatrix} a(z_1^{\circ}(\lambda), z_2^{\circ}(\lambda)) & b(z_1^{\circ}(\lambda), z_2^{\circ}(\lambda)) \\ c(z_1^{\circ}(\lambda), z_2^{\circ}(\lambda)) & d(z_1^{\circ}(\lambda), z_2^{\circ}(\lambda)) \end{vmatrix} \geq \delta > 0 \text{ on M.}$$

Then there exist $\varphi_1(x,y)$ and $\varphi_2(x,y)$ such that

(α) $\varphi_1(x,y)$ and $\varphi_2(x,y)$ are continuous in a region R: $x+y \geq 0$

(β) $z_1 = \varphi_1(x,y)$ and $z_2 = \varphi_2(x,y)$ are solutions of the system of equations:

$$(19.11) \qquad \begin{aligned} a(z_1, z_2) \frac{\partial z_1}{\partial x} + b(z_1, z_2) \frac{\partial z_2}{\partial x} &= f(z_1, z_2) \\ c(z_1, z_2) \frac{\partial z_1}{\partial y} + d(z_1, z_2) \frac{\partial z_2}{\partial y} &= g(z_1, z_2) \end{aligned}$$

(γ) $\varphi_j(\frac{\lambda}{\sqrt{2}}, \frac{-\lambda}{\sqrt{2}}) = z_j^{\circ}(\lambda)$ on M; that is, $\varphi_j(x,y) = z_j^{\circ}(\lambda)$ when (x,y) lies on L.

Proof: H. Lewy[1]. It is based upon method of finite differences: the (x,y) plane is covered by a square grid of mesh h: $x = \mu h$, $y = \nu h$, $(\mu,\nu = 0,\pm 1,\pm 2,\ldots)$ and 16.1 is replaced by the system of difference equations:

$$(19.12) \qquad \begin{aligned} a_{\mu\nu}(z_{1\mu+1,\nu} - z_{1\mu\nu}) + b_{\mu\nu}(z_{2\mu+1,\nu} - z_{2\mu\nu}) &= hf_{\mu\nu} \\ c_{\mu\nu}(z_{1\mu,\nu+1} - z_{1\mu\nu}) + d_{\mu\nu}(z_{2\mu,\nu+1} - z_{2\mu\nu}) &= hg_{\mu\nu} \end{aligned}$$

where $z_{1\mu\nu} \equiv \varphi_1(\mu h,\nu h)$ and $z_{2\mu\nu} = \varphi_2(\mu h,\nu h)$ and $a_{\mu\nu} = a(\mu h,\nu h),\ldots,$ $g_{\mu\nu} = g(\mu h,\nu h)$. To determine $z_{1\mu\nu}$ and $z_{2\mu\nu}$ to formally satisfy (β) and (γ): on the line L, we have lattice points $(\mu h, -\mu h)$, so by direct substitution in (γ),

$$z_{j\mu,-\mu} = \varphi_j(\mu h,-\mu h) = z_j^{\circ}(\sqrt{2}\mu h) \text{ for } \sqrt{2}\mu h \varepsilon M. \quad (j=1,2)$$

By equations (19.12) we can determine $z_{j\mu+1,\nu+1}$ if we know $z_{j,\mu,\nu+1}$

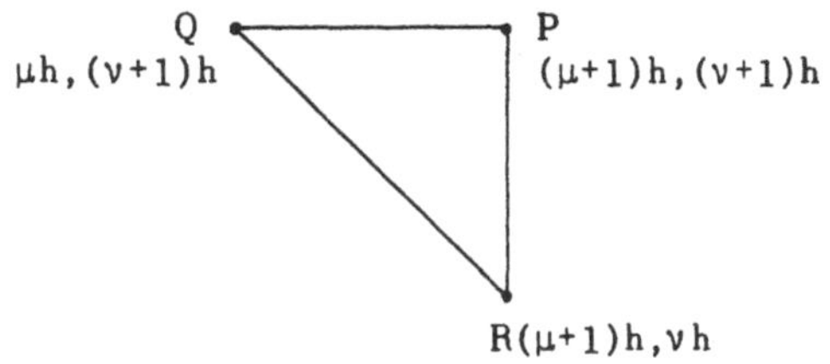

as is indicated in the above triangle, provided the determinant of coefficients is $\neq 0$. By (c) this is true within a certain distance of L; so we can determine first the values

$$z_{j\mu,-\mu+1} \text{ at points } (\mu h,(-\mu+1)h)$$

and then along succeeding lines parallel to L. Making the grid h as fine as we please, we get an approximate formula for $\varphi_j(x,y)$ - say $\bar{\varphi}_j^h(x,y)$, holding at points of the grid. As $h \to 0$, $\bar{\varphi}_j^h(x,y) \to \varphi_j(x,y)$ and their difference quotients remain uniformly continuous in any bounded region of the (x,y) plane, so that $\varphi_x(x,y) = \lim_{h \to 0} \frac{\bar{\varphi}_j^h(x+h,y)-\bar{\varphi}_j^h(x,y)}{h}$ exists and likewise $\varphi_y(x,y)$ exists.

This theorem was used by Lewy to establish the existence of solutions of the hyperbolic equation of second order, (see Ch. III). He generalized the result to systems of n equations in n unknowns, which are homogeneous. (Lewy[4])

THEOREM 19.5

(a) Let a_{ik} $(i=1,\ldots,n;\ k=1,\ldots,n)$ and b_i $(i=1,\ldots,n)$ be functions of class A^2 in the interior of a certain region S. in $(z_1, \ldots z_n)$-space

(b) Let line segment L be defined as in previous theorem.

(c) Let $z_0^i(\lambda)$ $(i=1,\ldots,n)$ be functions of class A^2 in M such that $(z_0^1(\lambda),\ldots,z_0^n(\lambda)) \varepsilon S$ and such that

$$|a_{ik}(z_0^1(\lambda),\ldots,z_0^n(\lambda))| \neq 0 \text{ for } \lambda \varepsilon M$$

Then there exist $\varphi^i(x,y)$ $(i=1,\ldots,n)$ such that

(α) $\varphi^i(x,y)$ $(i=1,\ldots,n)$ are of class A^1 in a region

R: $x+y \geq 0$

(β) $z_i = \varphi^i(x,y)$ $(i=1,\ldots,n)$ are solutions of the systems of equations:

$$\sum_{i=1}^{n} a_{ik} \frac{\partial z_k}{\partial x} = b_i \quad (i=1,\ldots,m);$$

$$\sum_{k=1}^{n} a_{ik} \frac{\partial z_k}{\partial y} = b_i \quad (i=m+1,\ldots,n)$$

(γ_o) When $(x,y)\ \varepsilon\ L$, $\varphi^i(x,y) \equiv \varphi^i\left(\frac{\lambda}{\sqrt{2}}, \frac{-\lambda}{\sqrt{2}}\right) = z_0^i(\lambda)$ $(i=1,\ldots,n)$

Furthermore φ^j are of class A^2, and the solution is unique (considering all solutions of class A^2).

Taylor [1] used the above theorem to obtain a solution $z_1 = \varphi_1(x,y)$, $z_1 = \varphi_2(x,y)$ of the system:

$$\begin{cases} F(x,y,z_1,z_2,p_1,p_2,q_1,q_2) = 0 \\ G(x,y,z_1,z_2,p_1,p_2,q_1,q_2) = 0 \end{cases}$$

such that on a given curve, z_i, $\frac{\partial z_i}{\partial x}$, $\frac{\partial z_i}{\partial y}$ $(i=1,2)$ take on prescribed values which satisfy strip equations. Again, one assumes this is done on a segment of $\alpha + \beta = 0$, so $x(t)$, $y(t)$, $\ldots$, $q_2(t)$ are functions of arc length on L.

Kourensky[1] considered the same system of equations, but his work is a generalization of the method of first integrals, and does not give an existence theorem. The general equation (19.1) was considered recently by Cinquini-Cibrario[1] in a paper which the author has not seen. Apparently she established the existence, in the neighborhood of the initial curve, of the solution to problem C, under the hypotheses that F^i be of class A^3 and satisfy certain Lipschitz conditions. The method of proof consists in reducing the problem to a quasi-linear system of a type considered in an earlier paper (Cinquini-Cibrario[2]):

$$(19.13)\quad \sum_{\kappa=1}^{r} a_{i\kappa}(x,y,z_1,\ldots z_r)\frac{\partial z_\kappa}{\partial x} = c_i(x,y,z_1,\ldots z_r)\quad [i=1,\ldots,m]\ ;$$

$$\sum_{\kappa=1}^{r} a_{i\kappa}\frac{\partial z_\kappa}{\partial y} = c_i\qquad [i=m+1,\ldots,r]$$

For this system, which is similar to Lewy's, existence of a solution is established when the initial data is given on any curve not parallel to either axis, instead of only the line $x+y = 0$. Her method is to differentiate the first group of equations with respect to x, the second group with respect to y, and then apply successive approximation of the Picard type.

Meltzer[1] considered the system (19.6) but instead of taking problem C or G, which assigns the values for $z_1, z_2, \ldots, z_r$ along the same curve C_0 in the (x,y) plane, he considered the problem called the "problem of Goursat" which assigns the values of some z_i on one curve and some on another. Specifically, he assumed that the coefficients in (19.6) are given in a bounded domain D, in the first quadrant, which has the segments L_1: $0\le x\le a$; $y = 0$ and L_2: $x=0$; $0\le y\le b$ as part of its boundary. He then obtained conditions under which there exists exactly one solution of (19.6) of class A^1 in D such that on L_1, $z_i = g_i(x)$ for $i = 1,\ldots,m$ while on L_2, $z_i = h_i(y)$ for $i=m+1,\ldots,r$, where g_i, h_i are any functions of class A^1, and also for which small changes in the initial values result in small changes in the solution. The conditions he obtained, however, are too complicated to be given here. Cinquini-Cibrario[2] considered a similar problem for (19.13).

20. Given a system of r equations in a single unknown function z:

$$(20.1)\quad F^{\nu}(x_1,\ldots,x_n,\ z,\ \frac{\partial z}{\partial x_1},\ldots,\frac{\partial z}{\partial x_n}) = 0 \qquad (\nu=1,\ldots,r;\ r<n)$$

a necessary condition that there exist a solution $z = \psi(x_1,\ldots,x_n)$ of class A^2 is that $\psi_{x_i x_j} = \psi_{x_j x_i}$, which means that

$$(20.2)\quad \sum_{j=1}^{n} \{[F^{\mu}_{x_j} + p_j F^{\mu}_z]F^{\nu}_{p_j} - [F^{\nu}_{x_j} + p_j F^{\nu}_z]F^{\mu}_{p_j}\} = 0$$

identically in $(x_1,\ldots,x_n)$ when $z = \psi$, $p_i = \psi_{x_i}$. If, given the functions $F^{\nu}(x_1,\ldots,x_n,z,p_1,\ldots,p_n)$, condition (20.2) is true, identically in x_i,z,p_i, then (20.1) is called an involutory system or completely integrable system; (20.2) is called the set of integrability conditions, and the left side is often abbreviated by: $[F^{\nu},\ F^{\mu}]$.

If the system $F^{\nu}(x_i,z,p_i) = 0$ can be solved for r of the variables x_i it is customary to denote these by $x_1,\ldots,x_r$ and the remaining variables by $y_1\ldots y_m$ $(m=n-r)$. Then we write the normal form of (20.1) as:

$$(20.3)\quad \frac{\partial z}{\partial x_k} = f^k(x_1,\ldots,x_r;\ y_1,\ldots,y_m;\ z;\ \frac{\partial z}{\partial y_1},\ldots,\frac{\partial z}{\partial y_n}) \quad (k = 1,\ldots,r)$$

The system is completely integrable, when

$$f^k(x_1,\ldots,x_r,\ y_1,\ldots,y_m,z,q_1,\ldots,q_m)$$

satisfies the identities:

$$(20.4)\quad \begin{aligned} &(f^k_{x_i} + f^k_z \cdot f^i) - (f^i_{x_k} + f^i_z \cdot f^k) \\ &+ \sum_{h=1}^{m} [f^k_{q_h}(f^i_{y_h} + f^i_z \cdot q_h) - f^i_{q_h}(f^k_{y_h} + f^k_z \cdot q_h)] \equiv 0 \end{aligned}$$

$$(i,k = 1,\ldots,r)$$

We give the theorems first for the analytic case and then when the functions are of any class A^m.

THEOREM 20.1

(a_∞) Let $(x_1^o,\dots,x_r^o)$ and $(y_1^o,\dots,y_m^o)$ be given, and let $g(y_1,\dots,y_m)$ be a function of class A^∞ in a neighborhood T_δ of $(y_1,\dots,y_m)$. Set $z^o = g(y_1^o,\dots,y_m^o)$ and $q_k^o = g_{y_k}(y_1^o,\dots,y_m^o)$.

(b_∞) Let $f^i(x_1,\dots,x_r;\ y_1,\dots,y_m;\ z;\ q_1,\dots,q_r)$ be of class A^∞ in a neighborhood S_δ of $(x_1^o,\dots,x_r^o;\ y_1^o,\dots,y_m^o;\ z^o;\ q_1^o,\dots,q_m^o)$.

(d) Let f^i satisfy (20.2) on page 79.

Then there exists a unique function $\Phi(x_1,\dots,x_r;\ y_1,\dots,y_m)$ such that:

(α_∞) $\Phi(x_1,\dots,x_r;\ y_1,\dots,y_m)$ is of class A^∞ in a neighborhood of R_δ of the point $(x_1^o,\dots,x_r^o;\ y_1^o,\dots,y_m^o)$.

(β_0) For all $(x_1,\dots,x_r;\ y_1,\dots,y_m)$ in R_δ. $z = \Phi(x_1,\dots,x_r;\ y_1,\dots,y_m)$ is a solution of the involutory, or completely integrable, or passive system:

$$(20.3)\qquad \frac{\partial z}{\partial x_k} = f^k(x_1,\dots,x_r;\ y_1,\dots,y_m;\ z;\ \frac{\partial z}{\partial y_1},\dots,\frac{\partial z}{\partial y_m})\ (k=1,\dots,r)$$

(γ_0) $\Phi(x_1^o,\dots,x_r^o;\ y_1,\dots,y_m) \equiv g(y_1,\dots,y_m)$ in T_{δ_1}.

Proof: One can establish the existence by the usual method of limits, and majorants, but Goursat[1], pp. 31-39 reduces the system to r equations of first order which are solved successively. (See the next theorem for another method of obtaining solutions.)

Germay[2] has again shown that after the system has been reduced to r first order equations, the method of successive approximations will yield the solution of the corresponding system of characteristic equations, and so on. (His notation is identical with Goursat so comparison is easy.)

THEOREM 20.2: (a_{m+1}) Let $(x_1^{\,o},\ldots,x_r^{\,o})$ be given and let $g(y_1,\ldots,y_n)$ be a function of class A^{m+1} $(m \geq 1)$ for all y.

(b_{m+1}) Let $f^i(x_1,\ldots,x_r;\ y_1,\ldots,y_n;\ z;\ q_1,\ldots,q_n)$ $(i=1,\ldots,r)$ be functions of class A^{m+1}, or more generally, let f^i, $f^i_{y_k}$, f^i_z, $f^i_{q_k}$ be functions of class A^m in $\overline{S}$:

$$|x_i - x_i^{\,o}| \leq a;\ -\infty < y_k, z, q_j < +\infty \qquad (i=1,\ldots,r;\ j,k=1,\ldots,n)$$

(d) Let f^i satisfy integrability conditions (19.4) on page 78.

(e) Let $|f^i_{x_\nu}|$, $|f^i_{y_k}|$, $|f^i_z|$, $|f^i_{q_k}|$, $|f^i_{y_i y_k}|$, $|f^i_{y_k z}|$, $|f^i_{y_k q_j}|$, $|f^i_{zz}|$, $|f^i_{z q_k}|$, $|f^i_{q_j q_k}|$ all be $\leq A$ in $\overline{S}_1$.

Let $0 < \beta < \frac{1}{rA} \log\left(1 + \frac{\log 3}{2n(B+1)}\right)$ and $\alpha = \min(a,\beta)$.

Then there exists a unique function $\varphi(x_1,\ldots,x_r,\ y_1,\ldots,y_n)$ such that

(α_2) $\varphi(x_1,\ldots,x_r;\ y_1,\ldots,y_n)$ is of class A^2 in $\overline{R}$:

$$|x_i - x_i^{\,o}| < \alpha,\ -\infty < y_k < +\infty \qquad (i=1,\ldots,r;\ k=1,\ldots,n)$$

(β_0) $z = \varphi(x_1,\ldots,x_r;\ y_1,\ldots,y_n)$ is a solution in $\overline{R}$ of

$$(20.3) \quad \frac{\partial z}{\partial x_k} = f^k\left(x_1,\ldots,x_r,\ y_1,\ldots,y_n,\ z,\ \frac{\partial z}{\partial y_1},\ldots,\frac{\partial z}{\partial y_n}\right)$$

$$(k=1,\ldots,r)$$

(γ_0) $\varphi(x_1^{\,o},\ldots,x_r^{\,o};\ y_1,\ldots,y_n) \equiv g(y_1,\ldots,y_n)$ for all y_i.

Furthermore $\varphi(x_1,\ldots,x_r,\ y_1,\ldots,y_n)$ is of class A^m in $\overline{R}$.

Proof: See Kamke[4]. Make the transformation:

$$(20.5) \quad x_i = x_i^{\,o} + u \cdot u_i \quad [i=1,\ldots,r].$$

That is, set

$$F(u,u_1,\ldots,u_r,y_1,\ldots y_m,Z,Q_1,\ldots Q_m) \equiv \sum_{\rho=1}^{r} f^\rho(x_1^o+uu_1,\ldots,x_r^o+uu_r,Z,Q_1,\ldots Q_m)$$

$$F^i(u,u_1,\ldots,u_r,y_1,\ldots y_m,Z,Q_1,\ldots Q_m) \equiv u\cdot f^i(x_1^o+uu_1,\ldots,x_r^o+uu_r,Z,Q_1,\ldots Q_m)$$

$$[i=1,\ldots r]$$

Find the solution of

$$\frac{\partial z}{\partial u} = F$$

such that

$$Z(0,u_1,..u_r,y_1,...y_m) = g(y_1,...y_m)$$

considering the u_i as parameters, by the methods of section 10. Call the solution

$$Z = \psi(u,u_1,...,u_r,y_1,...y_m)$$

(It will also satisfy $Z_{u_i} = F^i$ with $Z(u,0,...,0,y_j) = g(y_j)$). Then the solution to the given problem is shown to be:

$$z=\varphi(x_1,...,x_r;\ y_1,...y_m) \equiv \psi(1,x_1-\overset{o}{x}_1,...,x_r-\overset{o}{x}_r;\ y_1,...y_m)$$

In a footnote (p.269), Kamke says that his method is the method of Mayer transformations, found in Bieberbach[1] and Goursat[1] for the analytic case. The form of the Mayer transformations given in Goursat, pp.38-39 is:

$$x=u_1 + \overset{o}{x}_1;\ x_\kappa = \overset{o}{x}_\kappa + u_1 u_\kappa \quad (\kappa=2,...,r)$$

The system:

$$\frac{\partial Z}{\partial u_1} = f'(\overset{o}{x}_1+u_1,\overset{o}{x}_2+u_1u_2,...Z...) + \sum_{\kappa=2}^{r} u_\kappa \cdot f^\kappa$$

with $Z = g$ for $u_1 = 0$ is solved, and then $\varphi(x_1,...,x_r,y_1,...y_m)=\psi(x_1-\overset{o}{x}_1,\frac{x_2-\overset{o}{x}_2}{x_1-\overset{o}{x}_1}$ $,...,\frac{x_r-\overset{o}{x}_r}{x_1-\overset{o}{x}_1})$. This seems less useful than Kamke's form. In the same footnote, Kamke has some comments on the other two methods for solving involutory systems: the method of successive equations, mentioned in the last theorem, and the method of solving the characteristic equations of the given system. (See Caratheodory, p. 64).

As a corollary, we give the result for the simplest case - when $r=2$, and there are no parametric variables. The involutory system becomes:

$$(20.7) \qquad \frac{\partial z}{\partial x} = f^1(x,y,z);\ \frac{\partial z}{\partial y} = f^2(x,y,z)$$

and the integrability condition:

$$(20.8) \qquad f^1_y + f^1_z f^2 = f^2_x + f^2_z \cdot f^1$$

COROLLARY: Let f^i, f^i_z, be of class $A^k (k\geq 1)$ and let $f^i_x, f^i_y, f^i_z, f^i_{zz}$ be bounded $(i=1,2)$ and (20.8) be valid in the region: $|x-x_0|\leq a$; $|y-y_0|\leq a$; $-\infty<z<+\infty$. Let z_0 be a given constant. Then in the region $\bar{R}$: $|x-x_0|\leq a$; $|y-y_0|\leq a$ there exists exactly one solution $z=\varphi(x,y)$ of (20.7) which is of class A^1 in R, and such that $\varphi(x_0,y_0) = z_0$.

Proof: As in the theorem, find the solution $Z = \psi(u,u_1,u_2)$ of

$$\frac{\partial Z}{\partial u} = f^1(x_0+uu_1,y_0+uu_2,Z)\cdot u_1 + f^2(x_0+uu_1,y_0+uu_2,Z)\cdot u_2$$

such that $Z(0,u_1,u_2)=Z_0$, considering u_1 and u_2 as parameters. Then set $\varphi(x,y)=\psi(1,x-x_0,y-y_0)$. (See Bieberbach, p.276 for solution using Goursat's method.)

The lemma on page 45 will make the theorem applicable to finite regions. Finally, we have the corresponding result for linear involutory systems:

THEOREM 20.3

(a_m) Let $(x_1^0,\ldots,x_r^0)$ be any point in M: $a_i \le x_i \le b_i$ $(i=1,\ldots,r)$

Let $g(x,y_1,\ldots,y_n)$ be of class A^m for all y. $(m\ge 1)$

(b_m) Let $f^{\mu\sigma}(x_1,\ldots,x_r;\ y_1,\ldots,y_n)$ and $g^{\mu}(x_1,\ldots,x_r;\ y_1,\ldots,y_n)$ $(\mu=1,\ldots,r;\ \sigma=1,\ldots,n)$ be of class A^m $(m\ge 1)$ and let $f^{\mu\sigma}$ $(\mu=1,\ldots,r;\ \sigma=0,1,\ldots,n)$ be bounded in T: $a_i \le x_i \le b_i$; $-\infty < y_k < +\infty$.

(d) Let the integrability condition (20.4) hold. It becomes,

for the linear case,

$$\begin{cases} \sum_{\rho=1}^{n} (f^{\mu\sigma}_{y_\rho}\cdot f^{\nu\rho} - f^{\nu\sigma}_{y_\rho}\cdot f^{\mu\rho}) = f^{\mu\sigma}_{x_\nu} - f^{\nu\sigma}_{x_\mu} \\ \sum_{\rho=1}^{n} (g^{\mu}_{y_\rho}\cdot f^{\nu\rho} - g^{\nu}_{y_\rho}\cdot f^{\mu\rho}) = g^{\mu}_{x_\nu} - g^{\nu}_{x_\mu} + f^{\mu o}g^{\nu} - f^{\nu o}g^{\mu} \end{cases}$$

Then there exists a unique function $\varphi(x_1,\ldots,x_r;\ y_1,\ldots,y_n)$ such that;

(α_1) $\varphi(x_1,\ldots,x_r;\ y_1,\ldots,y_n)$ is of class A^m in T

(β_0) $z = \varphi(x_1,\ldots,x_r;\ y_1,\ldots,y_n)$ is a solution in T of the

system: $$\frac{\partial z}{\partial x_\mu} = \sum_{\sigma=1}^{n} f^{\mu\sigma}(x_1,\ldots,x_r;\ y_1,\ldots,y_n)\frac{\partial z}{\partial y_\sigma} + f^{\mu o}(x_1,\ldots,x_r;\ y_1,\ldots,y_n)z + g^{\mu}(x_1,\ldots,x_r;\ y_1,\ldots,y_n) \quad (\mu=1,\ldots,r)$$

(γ_0) $\varphi(x_1^0,\ldots,x_r^0;\ y_1,\ldots,y_n) \equiv g(y_1,\ldots,y_n)$ for all y.

Proof: Kamke [4], pp. 279-282.

The non-normal form of the homogeneous linear system:

$$\sum_{\kappa=1}^{n} f^{i\kappa}(x_1,\ldots,x_n)\frac{\partial z}{\partial x_\kappa} = 0 \qquad (i=1,\ldots,r)$$

has been discussed when it satisfies certain conditions which make it a " complete system" , Most of these discussions are not from the point of view of establishing existence theorems, and even in the definition of a solution of a complete system, there has been some ambiguity. Kamke[4] points out clearly what assumptions are involved concerning the successive reduction by transformations, where the matrices used are of rank $m<n$, and that these assumptions need not be true. Saltykow[1] discusses the question in connection with an investigation of just what is meant by the integrals of Lie for the system (20.1). Thus even when existence theorems are given for complete systems, such as that for the existence of principal integrals (Engel-Faber[1], p. 39) they should be carefully scrutinized, though some of them will probably still be true.

21. For systems of first order equations, m in number, and in r unknown functions, the work of Meray, Janet, and Riquier is fundamental; their considerations are true also for equations of any order and hence are found in chapter IV. Using the methods of Janet and Riquier, but with considerable simplification by tensor calculus notation, Thomas and Titt[1] gave conditions under which the general quasi-linear system

$$\sum_{\kappa=1}^{n} F^{j}_{i\kappa}\frac{\partial z_i}{\partial x_\kappa} = F^{j}_{0} \qquad (j=1,\ldots L)$$

of which (19.4) is a special case, can be reduced to the regular system:

$$\frac{\partial z_{i\kappa}}{\partial x^{\alpha}} = \Sigma\, G^{j}_{i\kappa}\, \frac{\partial z_{pq}}{\partial x^{\beta}} + \ldots .$$

where all the coefficients are of class A^{∞}, and hence, under suitable integrability conditions, when there exists an analytic solution of the problem of Cauchy.

III Second Order Differential Equations

A. Definitions, Classifications, Characteristic Equations, Transformations.

22. Let $F(x,y,z,p,q,r,s,t)$ be a function of class A^1 in a region U. If $\varphi(x,y)$ is such that when $(x,y)\ \varepsilon R$, $\varphi(x,y)$ is of class A^2 in R and $(x,y,\varphi,\varphi_x,\varphi_y,\varphi_{xx},\varphi_{xy},\varphi_{yy})\ \varepsilon U$ and

(22.1) $$F(x,y,\varphi_x(x,y),\varphi_y(x,y),\varphi_{xx}(x,y),\varphi_{xy}(x,y),\varphi_{yy}(x,y)) \equiv 0$$

then we say that $z = \varphi(x,y)$ is a *solution* of the *differential equation:*

(22.2) $$F(x,y,z,\frac{\partial z}{\partial x},\frac{\partial z}{\partial y},\ \frac{\partial^2 z}{\partial x^2},\frac{\partial^2 z}{\partial x \partial y},\frac{\partial^2 z}{\partial y^2}) = 0 \text{ in } R.$$

In discussing such equations, an important quantity is the discriminant

$$\Delta = 4RT - S^2$$

where

$$R = F_r(x,y,z,p,q,r,s,t)$$

$$S = F_s(x,y,z,p,q,r,s,t)$$

$$T = F_t(x,y,z,p,q,r,s,t).$$

If in a region $U_1 \leq U$

$$\begin{cases} \Delta > 0 \\ \Delta = 0 \\ \Delta < 0 \end{cases}$$

everywhere, equation (22.2) is called elliptic in U_1, parabolic in U_1, hyperbolic in U_1. It may happen that a given equation is in one of these catagories only locally - i.e. when U_1 is R_δ: $|x-x_0| < \delta$, $|y-y_0| < \delta$....

23. A quasi-linear equation of second order is obtained when

$$F(x,y,z,p,q,r,s,t) \equiv a(x,y) \cdot r + 2b(x,y) \cdot s + c(x,y) \cdot t + f(x,y,z,p,q)$$

so that (22.2) becomes:

$$(23.1) \quad a(x,y) \frac{\partial^2 z}{\partial x^2} + 2b(x,y) \frac{\partial^2 z}{\partial x \partial y} + c(x,y) \frac{\partial^2 z}{\partial y^2} + f(x,y,z, \frac{\partial z}{\partial x}, \frac{\partial z}{\partial y}) = 0.$$

Whether the equation is elliptic, parabolic, or hyperbolic will depend upon the behavior of a, b, c, in a region R_0 of the (x,y) plane, for

$$(23.2) \quad \Delta = 4[a(x,y) \cdot c(x,y) - (b(x,y))^2].$$

This is just the discriminant of the quadratic form:

$$(23.3) \quad a \cdot \xi^2 + 2b\xi\eta + c\eta^2$$

Hence we speak of (23.1) being elliptic, parabolic or hyperbolic in a region R_1 of the (x-y) plane. In many applications, $b(x,y) \equiv 0$, so that the equation (23.1) becomes:

$$(23.4) \quad a(x,y) \frac{\partial^2 z}{\partial x^2} + c(x,y) \frac{\partial^2 z}{\partial y^2} + f(x,y,z,\frac{\partial z}{\partial x},\frac{\partial z}{\partial y}) = 0$$

LEMMA 1: Given equation (23.4) where a(x,y) and c(x,y) are defined and continuous for $(x,y) \varepsilon R_0$ the equation is elliptic in $R_1 \leq R_0$ if and only if $a \neq 0$ everywhere in R_1, $c \neq 0$ everywhere in R_1 and a and c have the same sign; it is hyperbolic in $R_1 \leq R_0$ if and only if $a \neq 0$ everywhere in R_1, $c \neq 0$ everywhere in R_1 and a and c have opposite sign; it is parabolic in R_1 if $a \equiv 0$ or if $c \equiv 0$ in R_1. If in addition, it is required that a and c do not simultaneously vanish in R_1, then if $a \equiv 0$, $c \neq 0$ everywhere in R_1 and conversely.

Another important special case of (23.1) occurs when $a(x,y) = A$, $b(x,y) = B$, $c(x,y) = C$ and these are all constants, for then the classification of (23.1) is the same wherever $f(x,y,z,p,q)$ is defined. If this occurs for $(x,y) \in R_0$, and we assume A, B, C not all 0, we have:

LEMMA 2: Given equations

$$(23.5)\quad A\frac{\partial^2 z}{\partial x^2} + 2B\frac{\partial^2 z}{\partial x \partial y} + C\frac{\partial^2 z}{\partial y^2} + f(x,y,z,\frac{\partial z}{\partial x}, \frac{\partial z}{\partial y}) = 0$$

if $AC < 0$, (23.5) is hyperbolic; if $AC = 0$, $B \neq 0$, (23.5) is hyperbolic; if $AC = 0$, $B = 0$, (23.5) is parabolic; if $AC > 0$, $B = 0$, (23.5) is elliptic. (Hence, the only case where relative size of coefficients, as well as their signs, enters is when $AC > 0$, $B \neq 0$; if (23.5) is hyperbolic and $B = 0$, then $AC < 0$; if (23.5) is parabolic, then $AC \geq 0$; if (23.5) is parabolic and $B = 0$, then $AC = 0$; if (23.5) is parabolic and $AC = 0$, then $B = 0$.

In the general case of (23.1) one cannot make statements which are as definite as either lemma 1 or lemma 2. We can say that $a \cdot c < 0$ in R_1 $\to$ (23.1) is hyperbolic in R_1; $ac \leq 0$, $b \neq 0$ everywhere in $R_1 \to$ (23.1) is hyperbolic in R_1; conversely, if (23.1) is elliptic in R_1, then $ac > 0$ in R_1 and hence $a \neq 0$, $c \neq 0$ in R_1; if (23.1) is parabolic, $a \cdot c \geq 0$ everywhere in R_1; if (23.1) is parabolic and $a \cdot c \neq 0$ everywhere, then $b \neq 0$ everywhere; and if (23.1) is hyperbolic and $a \cdot c \geq 0$ everywhere, then $b \neq 0$ everywhere.

Another specialization of (23.1) would occur if

$$f(x,y,z,p,q) \equiv d(x,y) \cdot p + e(x,y) \cdot q + g(x,y) \cdot z - h(x,y);$$

the resulting equation is called a *linear equation of second order:*

$$(23.6)\quad a(x,y)\frac{\partial^2 z}{\partial x^2} + 2b(x,y)\frac{\partial^2 z}{\partial x \partial y} + c(x,y)\frac{\partial^2 z}{\partial y^2} + d(x,y)\frac{\partial z}{\partial x} + e(x,y)\frac{\partial z}{\partial y} + g(x,y)\, z = h(x,y).$$

If $h(x,y) \equiv 0$, it is homogeneous. Here again, the first three coefficients may be constant, or the first six coefficients may be constant, or all seven may be constant.

A type of equation which is not quasi-linear but which has been studied to a considerable extent is the Monge-Ampere equation which is (22.2) where

$$(23.7) \quad F \equiv \bar{a} + \bar{b}r + \bar{c}s + \bar{d}t + \bar{e}\,(rt-s^2)$$

and $\bar{a}$, $\bar{b}$, $\bar{c}$, $\bar{d}$, $\bar{e}$ are all functions of s, y, z, p, q. (When $\bar{e} = 0$, this equation is called, by some writers, a quasi-linear equation.)

$$(23.8) \quad \frac{\partial^2 z}{\partial x^2} = f(x,y,z,\frac{\partial z}{\partial x}, \frac{\partial z}{\partial y}, \frac{\partial^2 z}{\partial x \partial y}, \frac{\partial^2 z}{\partial y^2})$$

can be called a normal form of (22,2). A similar form is:

$$\frac{\partial^2 z}{\partial y^2} = f(x,y,z,\frac{\partial z}{\partial x}, \frac{\partial z}{\partial y}, \frac{\partial^2 z}{\partial x \partial y}, \frac{\partial^2 z}{\partial x^2})$$

Another kind of normal form would be:

$$(23.9) \quad \frac{\partial^2 z}{\partial x \partial y} = f(x,y,z,\frac{\partial z}{\partial x}, \frac{\partial z}{\partial y}, \frac{\partial^2 z}{\partial x^2}, \frac{\partial^2 z}{\partial y^2})$$

given (22.2) if F_r, F_s, F_t do not vanish simultaneously at any point of U, then in the neighborhood of a given point, (22.2) can be represented by one of these three **forms**.

24. We call a set of eight elements $(x_0,y_0,z_0,p_0,q_0,r_0,s_0,t_0)$ a *surface element of second order* and consider it as defining at each point (x_0,y_0,z_0) a paraboloid:

$$(24.1) \quad z-z_0 = p_0(x-x_0) + q_0(y-y_0) + \tfrac{1}{2}r_0(x-x_0)^2 + s_0(x-x_0)(y-y_0) + \tfrac{1}{2}t_0(y-y_0)^2$$

By a *strip of second order* we mean a totality of surface elements

$$(24.2) \quad x(\mu),\ y(\mu),\ z(\mu),\ p(\mu),\ q(\mu),\ r(\mu),\ s(\mu),\ t(\mu)$$

which are of class A^1 for $\mu \varepsilon M$ and satisfy certain equations which will insure that when the strip *lies on a surface* $z = \varphi(x,y)$, that is, when

$$(24.3)\quad \begin{aligned} z(\mu) &= \varphi(x(\mu),y(\mu)), \quad p(\mu) = \varphi_x(x(\mu),y(\mu)), \\ q(\mu) &= \varphi_y(x(\mu),y(\mu)) \\ r(\mu) &= \varphi_{xx}(x(\mu),y(\mu)), s(\mu) = \varphi_{xy}(x(\mu),y(\mu)), \\ t(\mu) &= \varphi_{yy}(x(\mu),y(\mu)) \end{aligned}$$

then we can differentiate twice continuously, and be consistent.

Thus from the first three, differentiating with respect to μ, (dropping μ in the symbols)

$$z' = \varphi_x \,.\, x' + \varphi_y \,.\, y' \text{ and hence } z' = p \cdot x' + q \cdot y'$$

$$p' = \varphi_{xx} \,.\, x' + \varphi_{xy} y' \text{ and hence } p' = r \cdot x' + s \cdot y'$$

$$q' = \varphi_{yx} \,.\, x' + \varphi_{yy} \,.\, y' \text{ and hence } q' = s \cdot x' + t \cdot y'$$

(provided φ is of class A^2).

Thus we can say a *strip of second order* is a set of elements (24.2) of class A^1 such that for all $\mu \varepsilon M$,

$$(24.4)\quad \begin{cases} z'(\mu) = p(\mu) \cdot x'(\mu) + q(\mu) \cdot y'(\mu) \\ p'(\mu) = r(\mu) \cdot x'(\mu) + s(\mu) \cdot y'(\mu) \\ q'(\mu) = s(\mu) \cdot x'(\mu) + t(\mu) \cdot y'(\mu). \end{cases}$$

By a procedure analogous to that used for first order equations one could set up conditions for *characteristic strips* which should be tangent to a *direction cone,* contain surface elements of second order, and have the direction of *initial strips.* This geometric reasoning is unnecessary for us here, so we simply write down the analytic definition (see Kamke[1], pp. 379-383 for derivation - or rather justification).

A set of elements (24.2) is called a *characteristic strip* of the differential equation

$$(22.2)\quad F(x,y,z,\frac{\partial z}{\partial x},\frac{\partial z}{\partial y},\frac{\partial^2 z}{\partial x^2},\frac{\partial^2 z}{\partial x \partial y},\frac{\partial^2 z}{\partial y^2}) = 0$$

if it is of class A^1 for $\mu\varepsilon M$, lies in the domain U of $F(x,y,z,p,q,r,s,t)$ for all $\mu\varepsilon M$, if $x = x(\mu)$, $y = y(\mu)$, $z = z(\mu), \ldots, t = t(\mu)$ constitute a solution of the six differential equations:

$$(24.5)\quad \begin{cases} z' = px' + qy' \\ p' = rx' + sy' \\ q' = sx' + ty' \\ Xx'^2 + Ss'x' + R(r'x' - s'y') = 0 \\ Yy'^2 + Ss'y' + T(t'y' - s'x') = 0 \\ Ry'^2 - Sx'y' + Tx'^2 = 0 \end{cases}$$

where

$$R \equiv F_r,\ S \equiv F_s,\ T \equiv F_t$$

$$X \equiv F_x + pF_z + rF_p + sF_q$$

$$Y \equiv F_y + qF_z + sF_p + F_q$$

and finally if $[x'(\mu)]^2 + [y'(\mu)]^2 > 0$ everywhere in M. Comparing the first three equations with (24.4) we see that a characteristic strip is a strip of second order.

An *integral strip* of equation (22.2) is a strip of second order (that is, (24.2) satisfying (24.4)) such that

$$(24.6)\qquad F(x(\mu),\ y(\mu),\ z(\mu), \ldots, t(\mu)) \equiv 0 \text{ in } M.$$

By (24.3) and (22.1) this means that $z = \varphi(x(\mu),\ y(\mu))$ will be a solution of (22.2). (An integral element of (22.2) is an element of second order such that $F(x_0, \ldots, t_0) = 0$.)

The two basic theorems are the following:

THEOREM 24.1: If $F(x,y,z,p,q,r,s,t)$ is of class A^1 in U, then along every characteristic strip of (22.2), $F(x,y,z,p,q,r,s,t)$ is constant. Hence, if a characteristic strip contains at least one integral element, it is an integral strip of (22.2).

THEOREM 24.2: If $F(x,\ldots,t)$ is of class A^1 in U and if $F^2 + R^2 + S^2 + T^2 > 0$ in U, and if $z = \varphi(x,y)$ is an integral of (22.2) lying in R_0 and of class A^3 there, and if $\Delta = 4RT - S^2 \leq 0$, for every surface element of second order on $z = \varphi(x,y)$, then through each integral element $(x_0,\ldots,t_0)$ of $z = \varphi(x,y)$, there is at least one characteristic strip containing this surface element (say for $\mu = \mu_0$) and belonging to the surface for a certain interval about μ_0. If $\Delta > 0$ for any surface element of second order on S: $z = \varphi(x,y)$, there is no real characteristic strip through this element.

Proof: We sketch the proof of theorem 24.2. Although it is no existence proof, it furnishes a useful method for obtaining characteristics (and also, of course, solutions in the neighborhood of a point). Let $\Delta \leq 0$.

For all (x,y,z) on $z = \varphi(x,y)$, $F(x,y,\ldots) = 0$ and hence $|R| + |S| + |T| > 0$ there. If $(x_0,y_0,\varphi(x_0,y_0),\ldots)$ is the given integral element, not all the four numbers

$$(24.7) \qquad 2R,\ S + \sqrt{S^2 - 4RT};\quad S - \sqrt{S^2 - 4RT},\quad 2T$$

are zero, and hence one or the other pair contains a term $\neq 0$; if this is the first pair, then in a neighborhood N of (x_0,y_0), $|2R| + |S + \sqrt{S^2 - 4RT}| > 0$, when we substitute $z = \varphi(x,y)$, $p = \varphi_x(x,y)$, etc..

Consider the system of equations:

$$(24.8) \qquad \begin{cases} x^{(1)} = 2R \\ y^{(1)} = S + \sqrt{S^2 - 4RT} \end{cases}$$

For some interval $a < \mu < b$, there is a solution $x = x(\mu)$, $y = y(\mu)$ such that $x_0 = x(\mu_0)$ and $y_0 = y(\mu_0)$. Also $|x'(\mu)| + |y'(\mu)| > 0$. Let $z(\mu) = \varphi(x(\mu),$

$y(\mu))$, etc., as in (24.3). Then we know this is a strip of second order, (24.4) and because $z = \varphi(x,y)$ was a solution, it satisfies (24.6) and therefore forms an integral element. It can be verified that it is a characteristic strip. When $\Delta > 0$, the last of equations (24.5) has no real solution and hence there is no real characteristic strip through such an an element.

If (22.2) is elliptic throughout U, there are no real characteristic strips, but if $\Delta \leq 0$ throughout U, and in particular if (22.2) is hyperbolic in U, one could try to form an integral surface out of characteristic strips starting from an initial strip, as was done for first order equations. (This procedure has not been carried through in general - see section B for remarks on its use.)

For the Monge-Ampere equation (23.7) the solution of the characteristic equations (24.5) can be simplified; for any characteristic integral strip (24.6), the first five functions also satisfy the following set of equations, which do not involve r, s, t:

$$(24.9)\quad \begin{cases} z' = px' + qy' \\ \bar{b}y'^2 - \bar{c}x'y' + \bar{d}x'^2 + \bar{e}(p'x' + q'y') = 0 \\ \bar{a}(x'^2 + y'^2) + (\bar{b}-\bar{d})(p'x' - q'y') + \bar{c}(p'y' \\ \quad + q'x') - \bar{c}(p'^2 + q'^2) = 0. \end{cases}$$

These are called the characteristic equations of (23.7). Any solution of them such that $x'^2 + y'^2 \; 0$ is called a characteristic strip of first order; for such a strip (24.4) and (24.6) can be used to determine the corresponding characteristic integral strip is necessary, but usually the first order strip will suffice. When $\bar{e} = 0$, the equations become:

$$(24.10)\quad z' = px' + qy';\ by'^2 - cx'y' + dx'^2 = 0;\ \bar{a}(x'^2 + y'^2) + (\bar{b}-\bar{d})(p'x' - q'y') + \bar{c}(p'y' + q'x') = 0$$

25. For the quasi-linear equation (23.1) the second of equations (24.10) becomes:

(25.1) $a \cdot y'^2 - 2\,bx'y' + cx'^2 = 0$

This will be called a characteristic equation of (23.1), since its solution can be found independently of the rest of the equations of the strip. A solution of (25.1) with

(25.2) $|x'(\mu)| + |y'(\mu)| > 0$

is called a characteristic curve C; it is just the projection on the (x-y) plane of the characteristic strip of first order (24.9) and indeed of the characteristic integral strip which is the solution of (24.5). If we need to obtain the remaining strip functions, we can use these equations, together with (23.1). We now consider how to find the characteristic curve through any point and in a region.

Let λ_1 and λ_2 be the roots of the algebraic equation:

(25.3) $\lambda^2 + 2b\lambda + ac = 0$

which we can form for any point (x,y) in R_0. Then

(25.4) $\lambda_1 = -b(x,y) + \sqrt{b^1(x,y) - a(x,y)\cdot c(x,y)}$

$\lambda_2 = -b(x,y) - \sqrt{b^2(x,y) - a(x,y)\cdot c(x,y)}$

We can think of these two equations as defining functions $\lambda_i(x,y)$ throughout R, whether or not the values of these functions are real or distinct. They are related to the functions a(x,y), b(x,y), c(x,y) by the identities:

(25.5) $\lambda_1+\lambda_2 = -2b$; $\lambda_1\lambda_2 = ac$

(When $\Delta = b^2-ac\geq 0$ throughout R, $\sqrt{\Delta}$ will always mean the non-negative square-root; if complex values are also considered, $\sqrt{\Delta}$ will be the root whose amplitude θ satisfies: $-\frac{\pi}{2} < \theta \leq \frac{\pi}{2}$.)

THEOREM 25.1: Let a, b, c be continuous, $ac \leq b^2$, and $|a| + |b| + |c| > 0$ everywhere in R_0. Then through each point of R_0 there passes at least one characteristic curve

$$C_0 \begin{cases} x = x(\mu) \\ y = y(\mu) \end{cases}$$

of the quasi-linear equation:

$$(23.1) \quad a(x,y)\frac{\partial^2 z}{\partial x^2} + 2b(x,y)\frac{\partial^2 z}{\partial x \partial y} + c(x,y)\frac{\partial^2 z}{\partial y^2} + f(x,y,z,p,q) = 0.$$

If P^0 is a point at which $ac < b^2$, there are exactly two characteristic curves through P^0, which are not tangent to each other.

Proof: Under the conditions of the theorem, each $\lambda_i(x,y)$ is real and continuous in R_0. Let P^0 be a given point (x_0, y_0) and suppose, first of all, that $\lambda_1 \neq 0$ at P^0. Then there is a unique solution of

$$(25.7) \quad x' = a(x,y) \ ; \ y' = -\lambda_1(x,y)$$

which passes through P^0 for $\mu = 0$, since $a^2 + \lambda_1{}^2 > 0$ for a neighborhood of P^0. This solution: $x = \varphi(\mu)$; $y = \psi(\mu)$ satisfies (25.2) and also $ay'^2 - 2bx'y' + cx'^2 = a\lambda_1{}^2 + 2ab\lambda_1 + ca^2 = 0$ so that it is a characteristic curve Γ_1. It can also be represented, non-parametrically, as the solution of

$$(25.8) \quad \frac{dx}{dy} = -\frac{a(x,y)}{\lambda_1(x,y)}$$

which goes through (x_0, y_0); in this form, we denote it by $x = A(y; x_0, y_0)$. There is also a unique solution of

$$(25.9) \quad x' = -\lambda_1(x,y) \ ; \ y' = c(x,y)$$

passing through P^0 for $\nu = 0$; writing it $x = \overline{\varphi}(\nu)$; $y = \overline{\psi}(\nu)$, it satisfies (25.2) and also $ay'^2 - 2bx'y' + cx'^2 = ac^2 + 2bc\lambda_1 + c\lambda_1{}^2 = 0$, so that it is a characteristic curve Γ_2. Non-parametrically, it can be represented as a solution of

$$(25.10) \quad \frac{dy}{dx} = -\frac{c(x,y)}{\lambda_1(x,y)}$$

which goes through (x_0, y_0); in this form, we denote it by $y = B(x; x_0, y_0)$. If $ac<b^2$, curves Γ_1 and Γ_2 are not tangent to each other at P^0 since if they were, either $c \neq 0$ and $\frac{a}{-\lambda_1} = -\frac{\lambda_1}{c}$ from (25.7) and (25.9) which means $\lambda_1{}^2 = ac = \lambda_1\lambda_2$ by (25.5) which is impossible since $\lambda_1 \neq \lambda_2$; or else $c=0$, whence $0 = -\frac{\lambda_1}{a}$, which is a contradiction since $\lambda_1 \neq 0$.

If $\lambda_2 \neq 0$ at P^0, we could carry through the same process, finding a unique solution $x = \lambda(\mu)$, $y = \theta(\mu)$ of:

$$(25.11) \quad x' = a(x,y), \; y' = -\lambda_2(x,y)$$

through (x_0, y_0) for $\mu = 0$. This turns out to be a characteristic curve $\overline{\Gamma}_2$ for some neighborhood of P^0, which can be represented non-parametrically as $x = \overline{B}(y; x_0, y_0)$, the solution of

$$(25.12) \quad \frac{dx}{dy} = -\frac{a(x,y)}{\lambda_2(x,y)}$$

and a unique solution of

$$(25.13) \quad x' = -\lambda_2(x,y); \; y' = c(x,y)$$

through (x_0, y_0) for $\nu = 0$. This is also a characteristic curve $\overline{\Gamma}_1$, for some neighborhood of P^0 and can be represented non-parametrically: $y = \overline{A}(x; x_0, y_0)$, the solution through P^0 of

$$(25.14) \quad \frac{dy}{dx} = -\frac{c(x,y)}{\lambda_2(x,y)}$$

If $ac<b^2$ these two curves are not tangent to each other.

Now if at the same point P^0, both λ_1 and λ_2 were not 0, there would be two sets of two equations each. But from (25.5), $\lambda_1\lambda_2 = ac \neq 0$, so neither a nor c vanishes at P^0 and one can easily verify that (25.10) and (25.12) are equivalent, since $\frac{a}{\lambda_2} = \frac{\lambda_1}{c}$. Thus Γ_2 and $\overline{\Gamma}_2$ are solutions of the same equations through the same point and so are identical curves. Likewise, (25.8) and (25.14) represent the same characteristic curve.

If $ac<b^2$ at P^0 either λ_1 or λ_2 is different from 0, so at least one of the two groups above can be used, and there are always two characteristic curves through P^0. If $ac = b^2$, then $\lambda_1 = \lambda_2$. If their common value is not 0, we still have (25.7) or (25.8) and at least one characteristic curve through P^0. If $\lambda_1 = \lambda_2 = 0$, then $b = 0$ but either a or c is not 0 at P^0 and so for some neighborhood of P^0. One of the equations (25.7) and (25.9) has a solution, which is a characteristic curve for some neighborhood of P^0.

COROLLARY 1 If (23.1) is hyperbolic throughout R, there are exactly two characteristics through every point of R. When $\lambda_1 \neq 0$ throughout R, they can be written:

(25.15) $x = A(y;x_0,y_0)$; $y = B(x;x_0,y_0)$

where these are the solutions of (25.8) and (25.10) respectively; each of these functions is of class A^1 in all three variables. If the equations were solved, respectively, for x_0 and y_0, the result would be: $x_0 = A(y_0;x,y)$, $y_0 = B(x_0;x,y)$. No curve of either family is tangent to a curve of the other, at any point.

By using (25.7) and (25.11), we get another result which is sometimes useful, especially for equations in normal form:

COROLLARY 2 When $a \neq 0$ throughout R, the characteristics through (x_0,y_0) are

$$y = D(x;\ x_0,y_0)\ ;\ y = E(x;\ x_0,y_0)$$

where the first is a solution of $\frac{dy}{dx} = -\frac{\lambda_1}{a}$ and the second a solution of $\frac{dy}{dx} = -\frac{\lambda_2}{a}$. Each function D, E is of class A^1; solving the equations gives: $y_0 = \overline{D}(x_0;x,y)$; $x_0 = \overline{E}(y_0;x,y)$. If (23.1) is hyperbolic throughout R, the two systems are distinct; if (23.1) is parabolic throughout R, they coincide.

For the special forms mentioned in section 23, the equations of the characteristics becomes simple. For example, if (23.4) is hyperbolic, then it has $a \cdot c < 0$ everywhere in R; the two distinct characteristics through (x_0, y_0) are given everywhere by the solutions of:

$$y' \pm k(x,y)\, x' = 0$$

where $k(x,y) = \sqrt{-\frac{c(x,y)}{a(x,y)}}$ which pass through (x_0, y_0), and have $|x'| + |y'| > 0$.

THEOREM 25.2: When A, B, C are constants the characteristics of

$$(23.5)\quad A\frac{\partial^2 z}{\partial x^2} + 2B\frac{\partial^2 z}{\partial x \partial y} + C\frac{\partial^2 z}{\partial y^2} + f(x,y,z,\frac{\partial z}{\partial x},\frac{\partial z}{\partial y}) = 0 \quad [B \leq 0]$$

are straight lines defined throughout the plane. If (23.5) is hyperbolic, and we define $\lambda_1 = -B + \sqrt{B^2 - AC}$, then the characteristics through (x_0, y_0) are

$$y - y_0 = \frac{C}{-\lambda_1}(x - x_0)\ ; \qquad x - x_0 = \frac{A}{-\lambda_1}(y - y_0)$$

If (23.5) is parabolic there is a single characteristic through (x_0, y_0); it is:

$$y - y_0 = \frac{B}{A}(x - x_0) \text{ if } A \neq 0;$$

or

$$x - x_0 = \frac{B}{C}(y - y_0) \text{ if } C \neq 0.$$

When the general equation (23.1) is given, it is possible to reduce it to a form where characteristics are straight lines. For instance, if it is hyperbolic, and if $\lambda_1 \neq 0$, the functions $U(x,y) = A(x_0; x,y)$; $V(x,y) = B(x_0; x,y)$ by corollary 1, would effect the reduction. But one has to be careful, since the mere fact that either λ_1 or λ_2 is not 0 at every point of R does not mean that one of them does not vanish throughout R and hence that equations such as (25.15) hold throughout R. In 1936, Kamke[3] did prove that such a transformation was possible:

THEOREM 25.3: Let $a(x,y)$, $b(x,y)$, $c(x,y)$ be of class A^{m+2} $(m \geq 0)$ in a simply connected region G of the (x,y) plane and let $ac < b^2$ everywhere in G. Let R be any bounded region in the interior of G. Let $\varphi(x,y)$ be a function of class A^{m+2} in R such that $f(x,y,z,p,q)$ is defined whenever $(x,y) \in R$, $z = \varphi$, $p = \varphi_x$, $q = \varphi_y$.

Then there exists two functions $U(x,y)$ and $V(x,y)$ of class A^{m+2} in R such that the transformation:

$$T: \quad \xi = U(x,y), \quad \eta = V(x,y)$$

has Jacobian

$$(25.16) \qquad J = \begin{vmatrix} U_x & U_y \\ V_x & V_y \end{vmatrix} > 0 \text{ in } R$$

and maps R in a 1-1 manner onto a region $\bar{R}$ of the (ξ,η) plane; the inverse transformation:

$$T^{-1}: \quad x = X(\xi,\eta), \quad y = Y(\xi,\eta)$$

when applied to φ gives a function:

$$\bar{\varphi}(\xi,\eta) \equiv \varphi(X(\xi,\eta)\, Y(\xi,\eta))$$

of class A^{m+2} in $\bar{R}$. If we replace x and y by their values from T^{-1} in:

$$(25.17) \quad a(x,y)\varphi_{xx}(x,y) + 2b(x,y) \cdot \varphi_{xy}(x,y) + c(x,y) \cdot \varphi_{yy}(x,y) + f(x,y,\varphi,\varphi_x,\varphi_y)$$

it becomes:

$$(25.18) \quad 2B(\xi,\eta) \cdot \bar{\varphi}_{\xi\eta}(\xi,\eta) + D(\xi,\eta) \cdot \bar{\varphi}_{\xi}(\xi,\eta) + E(\xi,\eta) \cdot \bar{\varphi}_{\eta}(\xi,\eta) + f[X,Y,\bar{\varphi},\bar{\varphi}_{\xi}U_x + \bar{\varphi}_{\eta}V_x,\bar{\varphi}_{\xi}U_y + \bar{\varphi}_{\eta}V_y]$$

where B, D, E are functions of class A^m and $B(\xi,\eta) \neq 0$ in $\bar{R}$, and where $U_x \equiv U_x[X(\xi,\eta),Y(\xi,\eta)]$, etc.

Proof: Kamke[3] pp. 297-299. Let $\lambda_1 = -b + \sqrt{b^2-ac}$ and $\lambda_2 = -b-\sqrt{b^2-ac}$. They are functions of class A^{m+2} in G and $\lambda_1\lambda_2 = ac$, so that $ac - \lambda_1\lambda_2 = 0$ in G.

But then the matrices $\begin{Vmatrix} a & -\lambda_2 \\ -\lambda_1 & c \end{Vmatrix}$ and $\begin{Vmatrix} a & -\lambda_1 \\ -\lambda_2 & c \end{Vmatrix}$ are both of rank 1, and hence using a lemma Kamke proved at the beginning of his paper, there are two functions $\alpha_1(x,y)$ and $\alpha_2(x,y)$ of class A^m in G such that, for suitably chosen $\rho_\nu(x,y)$, $\sigma_\nu(x,y)$ $(\nu = 1,2)$, and with $f_\nu = \cos \alpha_2$ and $g_\nu = \sin \alpha_\nu$, we have:

$$(25.19) \quad a = \rho_1 f_1, \ -\lambda_2 = \sigma_1 f_1, \ -\lambda_1 = \rho_1 g_1, \ c = \sigma_1 g_1$$

and

$$(25.20) \quad a = \rho_2 f_2, \ -\lambda_2 = \rho_2 g_2, \ -\lambda_1 = \sigma_2 f_2, \ c = \sigma_2 g_2.$$

Now $f_1 g_2 - f_2 g_1 \neq 0$ $[\cos \alpha_1 \sin \alpha_2 - \cos \alpha_2 \sin \alpha_1 = \sin(\alpha_1 - \alpha_2)]$ and hence

$$\begin{vmatrix} \rho_1 & \rho_2 \\ \sigma_1 & \sigma_2 \end{vmatrix} \cdot \begin{vmatrix} f_1 & g_1 \\ f_2 & g_2 \end{vmatrix} = \begin{vmatrix} \rho_1 f_1 + \rho_2 f_2 & \rho_1 g_1 + \rho_2 g_2 \\ \sigma_1 f_1 + \sigma_2 f_2 & \sigma_1 g_1 + \sigma_2 g_2 \end{vmatrix} = \begin{vmatrix} 2a & -(\lambda_1+\lambda_2) \\ -(\lambda_1+\lambda_2) & 2c \end{vmatrix}$$

$$= 4ac - 4b^2 < 0$$

But from this, one can (if necessary, replacing f_2 by $-f_2$, g_2 by $-g_2$, ρ_2 by $-\rho_2$, σ_2 by $-\sigma_2$), assume that $f_1 g_2 - f_2 g_1 > 0$ in G.

Apply theorem 19.3 to the equations:

$$(25.21) \quad f_\nu(x,y) \frac{\partial z_\nu}{\partial x} + g_\nu(x,y) \frac{\partial z_\nu}{\partial y} = 0 \qquad (\nu = 1,2)$$

and we find there are two functions U(x,y) and V(x,y) such that $z_1 = U(x,y)$ and $z_2 = V(x,y)$ are solutions of these equations in R, and are of class A^{m+2} in R, and have $J > 0$ in R. Hence the transformation T will map R onto $\bar{R}$ as stated, and inverse functions are of class A^{m+2}.

Also, since $\varphi(x,y) \equiv \bar{\varphi}(U(x,y), V(x,y))$

$$\varphi_x = \bar{\varphi}_\xi(U(x,y), V(x,y)) \cdot U_x(x,y) + \bar{\varphi}_\eta(U(x,y), V(x,y)) \cdot V_x(x,y)$$

and similarly (arguments on right side are same as in above line),

$$\varphi_y = \bar{\varphi}_\xi \cdot U_y + \bar{\varphi}_\eta \cdot V_y$$

$$\varphi_{xx} = \bar{\varphi}_{\xi\xi} \cdot U_x^2 + 2\bar{\varphi}_{\xi\eta} \cdot U_x V_x + \bar{\varphi}_{\eta\eta} \cdot V_x^2 + \varphi_\xi \cdot U_{xx} + \bar{\varphi}_\eta \cdot V_{xx}$$

$$\varphi_{xy} = \bar{\varphi}_{\xi\xi} \cdot U_x U_y + \bar{\varphi}_{\xi\eta}(U_x V_y + V_x U_y) + \bar{\varphi}_{\eta\eta} \cdot V_x V_y + \bar{\varphi}_\xi \cdot U_{xy} + \bar{\varphi}_\eta \cdot V_{xy}$$

$$\varphi_{yy} = \bar{\varphi}_{\xi\xi} \cdot U_y^2 + 2\bar{\varphi}_{\xi\eta} U_y V_y + \bar{\varphi}_{\eta\eta} \cdot V_y^2 + \bar{\varphi}_\xi \cdot U_{yy} + \bar{\varphi}_\eta V_{yy}$$

If we substitute in(25.17)and collect coefficients and then use T, no matter what the transformation is, provided $J \neq 0$, we get:

$$(25.22) \quad \alpha \cdot \bar{\varphi}_{\xi\xi} + 2\beta \cdot \bar{\varphi}_{\xi\eta} + \gamma\bar{\varphi}_{\eta\eta} + \delta\bar{\varphi}_\xi + \varepsilon\bar{\varphi}_\eta + \zeta$$

where $\alpha \equiv [aU_x^2 + 2bU_xU_y + cU_y^2]$ with $x = X(\xi,\eta)$, $y = Y(\xi,\eta)$

$\beta \equiv [aU_xV_x + b(U_xV_y + V_xU_y) + cU_yV_y]$ "

$\gamma \equiv [aV_x^2 + 2bV_xV_y + cV_y^2]$ "

$\delta \equiv [aU_{xx} + 2bU_{xy} + cU_{yy}]$ $\quad \varepsilon \equiv [aV_{xx} + 2bV_{xy} + cV_{yy}]$

and $\zeta \equiv \bar{f}(\xi,\eta,\bar{\varphi},\bar{\varphi}_\xi,\bar{\varphi}_\eta)$ with $\bar{f}(\xi,\eta,\omega,\bar{p},\bar{q}) \equiv f(x,y,\omega,p,q)$ where $x = X(\xi,\eta)$, $y = Y(\xi,\eta)$, $p = \bar{p} \cdot U_x + \bar{q}\, V_x$ +and $q = P \cdot U_y + \bar{q}\, V_y$, are substituted on the right hand side. Note that

$$\bar{p} = \frac{p\,V_y - q\,V_x}{J} = p\,X_\xi + q\,Y_\xi \qquad \text{since} \qquad \begin{cases} X_\xi = \frac{+V_y}{J}, & X_\eta = \frac{-U_y}{J} \\ Y_\xi = \frac{-V_x}{J}, & Y_\eta = \frac{U_x}{J}. \end{cases}$$

$$\bar{q} = \frac{q\,U_x - pU_y}{J} = p\,X_\eta + q\,Y_\eta$$

Now with the particular $U(x,y)$ and $V(x,y)$ defined by(25.19) (25.20) and (25.21), the following identities hold everywhere in R:

$$(*) \begin{cases} aU_x - \lambda_1 U_y \equiv \rho_1 f_1 U_x + \rho_1 g_1 U_y \equiv \rho_1(f_1 U_x + g_1 U_y) \equiv 0 & \therefore\ aU_x \equiv \lambda_1 U_y \\ -\lambda_2 U_x + cU_y \equiv \sigma_1(f_1 U_x + g_1 U_y) \equiv 0 & \therefore\ cU_y \equiv \lambda_2 U_x \\ aV_x - \lambda_2 V_y \equiv \rho_2(f_2 V_x + g_2 V_y) \equiv 0 & \therefore\ aV_x \equiv \lambda_2 V_y \\ -\lambda_1 V_x + cV_y \equiv \sigma_2(f_2 V_x + g_2 V_y) \equiv 0 & \therefore\ cV_y \equiv \lambda_1 V_x \end{cases}$$

Hence $\alpha \equiv U_x \cdot \lambda_1 V_y + 2bU_xU_y + U_y \cdot \lambda_2 U_x \equiv U_xU_y\ [2b + \lambda_1 + \lambda_2] \equiv 0$

$\gamma \equiv \lambda_2 V_yV_x + 2bV_xV_y + V_y \cdot \lambda_1 V_x \equiv V_xV_y\ [2b + \lambda_1 + \lambda_2] \equiv 0$

$\beta \equiv \lambda_1 U_yV_x + b(U_xV_y + V_xU_y) + \lambda_2 U_xV_y \equiv U_yV_x(b+\lambda_1) + U_xV_y(b+\lambda_2)$

$\equiv -\sqrt{b^2 - ac} \cdot J(x,y) \neq 0$ everywhere.

Under T, therefore, $\beta = -\sqrt{b^2(X(\xi,\eta) \cdot Y(\xi,\eta)) - a(X,Y) \cdot c(X,Y)} \cdot \frac{1}{J(\xi,\eta)} \neq 0$

So we get (25.8).

COROLLARY: Under transformation T, the characteristics through a point (x_0, y_0), go into the lines $\xi = \xi_0$ and $\eta = \eta_0$ where $\xi_0 = U(x_0, y_0)$ and $\eta_0 = V(x_0, y_0)$. The curve Γ_1 which is the solution of (25.7) p. 94, goes into $\xi = \xi_0$ (for some range of values of η including η_0) and the curve Γ_2, which is the solution of (25.9) p. 94, goes into $\eta = \eta_0$ (for some range of values of ξ, including ξ_0), provided $\lambda_1 \neq 0$ at (x_0, y_0). [If $\lambda_1 = 0$, $\lambda_2 \neq 0$, use $\overline{\Gamma}_1$ and $\overline{\Gamma}_2$ respectively].

Proof: By (*), $z = U(x,y)$ is a solution of: $a(x,y)\frac{\partial z}{\partial x} - \lambda_1(x,y)\frac{\partial z}{\partial y} = 0$
When $\lambda_1 \neq 0$, Γ_1 is: $x = x_0(\mu)$, $y = y_0(\mu)$, where these are the solutions of $x' = a(x,y)$, $y' = -\lambda_1(x,y)$ and hence is a characteristic curve of the above equation, in the sense used on page 52. By theorem 14.1, U is constant along Γ_1 - that is, $U[x_0(\mu), y_0(\mu)] = \text{const.}$ But $U[x_0(0), y_0(0)] = U[x_0, y_0] = \xi_0$ and hence $U[x_0(\mu), y_0(\mu)] = \xi_0$. Also $V[x_0(\mu), y_0(\mu)] = h(\mu)$, where $h'(\mu) \neq 0$, by (*). Hence under T, a point of Γ_1 goes into (ξ_0, η) for some interval containing (ξ_0, η_0). A similar argument applies for Γ_2, and for the case $\lambda_2 \neq 0$.

THEOREM 25.4: In the preceding theorem, let the hypothesis that $b^2 - ac > 0$ be replaced by:

$$b^2 - ac \equiv 0 \text{ in } G; \quad |a| + |c| > 0 \text{ in } G.$$

Then the result of theorem 25.3 is still valid except that (25.18) is replaced by:

$$(25.23)\quad C(\xi,\eta)\cdot\bar{\varphi}_{\eta\eta}(x,y)+D(\xi,\eta)\cdot\bar{\varphi}_{\xi}(\xi,\eta)+E(\xi,\eta)\cdot\bar{\varphi}_{\eta}(\xi,\eta)+f[X(\xi,\eta),Y(\xi,\eta),\bar{\varphi}(\xi,\eta),\bar{\varphi}_{\xi}\cdot U_x+\bar{\varphi}_{\eta}\cdot V_x,\ \bar{\varphi}_{\xi}\cdot U_y+\bar{\varphi}_{\eta}\cdot V_y]$$

where $C \neq 0$, and C, D, E are of class A^m. Kamke [3], p. 299-300.

Proof: Defining λ_1 and λ_2 as on page 98, $\lambda_1 = \lambda_2 = -b$, so the matrices both reduce to $\begin{Vmatrix} a & b \\ b & c \end{Vmatrix}$, which is of rank 1. But by the lemma, again, there are in G functions f_1 and g_1 of class A^{m+1} with $f_1^2 + g_1^2 > 0$, so that for certain ρ and σ,

$$(25.24)\quad a = \rho f_1;\quad c = \sigma g_1;\quad b = \rho g_1 = \sigma f_1$$

Setting $f_2 = -g$, and $g_2 = f_1$, this means that $f_1 g_2 - f_2 g_1 = f_1^2 + g_1^2 > 0$ and so by the same theorem used before, we get solutions of (25.21), which now become:

$$(25.25)\quad \begin{cases} f_1(x,y)\dfrac{\partial z_1}{\partial x} + g_1(x,y)\dfrac{\partial z_1}{\partial y} = 0 \\ -g_1(x,y)\dfrac{\partial z_2}{\partial x} + f_1(x,y)\dfrac{\partial z_2}{\partial y} = 0 \end{cases}$$

The solutions U(x,y), V(x,y) have required properties and transformations T and T^{-1} are defined as before, resulting in reduction of (25.17) to (25.22) again. But now from (25.24),

$$aU_x + bU_y \equiv \rho f_1 U_x + \rho g_1 U_y \equiv 0$$

$$+\ bU_x + cU_y \equiv \sigma f_1 U_x + \sigma g_1 U_y \equiv 0$$

$$-\ bV_x + aV_y \equiv -\rho g_1 V_x + \rho f_1 V_y \equiv 0$$

$$-\ cV_x + bV_y \equiv -\sigma g_1 V_x + \sigma f_1 V_y \equiv 0$$

$$\therefore\ \alpha \equiv U_x(-bU_y) + 2bU_xU_y + U_y(-bU_x) \equiv U_xU_y(2b-2b) \equiv 0$$

$$\beta \equiv (aU_x + bU_y)\cdot V_x + (bU_x + cU_y)\,V_y \equiv 0\cdot V_x + 0\cdot V_y \equiv 0$$

$$\gamma \equiv [aV_x^2 + aV_y^2 + cV_x^2 + cV_y^2] \equiv 2(a+c)\ (V_x^2+V_y^2)$$

But $V_x^2 + V_y^2 \neq 0$, and since $b^2 - ac = 0$, $ac \geq 0$, so a and c have the same sign, and $a+c = \pm(|a| + |c|) \neq 0$, so $\gamma \neq 0$.

Thus (25.22) becomes (25.23).

For the elliptic case, there is no analogy to theorem 25.2, since the characteristic equation (24.10) has only imaginary factors. Nevertheless there is a theorem analogous to theorem 25.3. Since $ac-b^2 > 0$ in R, $a \neq 0$ and $c \neq 0$ anywhere in R.

THEOREM 25.5 Let the hypotheses of theorem 25.3 hold, except that the condition that $ac < b^2$ everywhere in G is replaced by: $ac > b^2$ everywhere in G. Then the theorem is still valid, except that (25.18) is replaced by

$$(25.26)\quad A(\xi,\eta) \cdot \bar{\varphi}_{\xi\xi}(\xi,\eta) + C(\xi,\eta) \cdot \bar{\varphi}_{\eta\eta}(\xi,\eta,\ldots + D(\xi,\eta) \cdot \bar{\varphi}_{\xi}(\xi,\eta) + E(\xi,\eta) \cdot \bar{\varphi}_{\eta}(\xi,\eta) + \bar{f}(\xi,\eta,\ldots\ )$$

where all coefficients are functions of class A^m and $A(\xi,\eta) > 0$ and $C(\xi,\eta) > 0$ everywhere in $\bar{R}$.

Proof: Since $ac - b^2 > 0$, if we set $f_1(x,y) = a(x,y)$, $f_2(x,y) = g_1(x,y) = b(x,y)$, $g_2(x,y) = c(x,y)$, the equations (25.21) will have $f_1 g_2 - f_2 g_1 = ac - b^2 > 0$ in R. Hence by the corollary to theorem 19.3, p. 74, there exist functions U(x,y) and V(x,y) such that $z_1 = U(x,y)$ and $z_2 = V(x,y)$ are solutions of (25.21), which becomes:

$$(25.27)\quad \begin{cases} a(x,y)\dfrac{\partial z_1}{\partial x} + b(x,y)\dfrac{\partial z_1}{\partial y} = 0 \\[2ex] b(x,y)\dfrac{\partial z_2}{\partial x} + c(x,y)\dfrac{\partial z_2}{\partial y} = 0 \end{cases}$$

which are of class A^{m+2} in R and have $J = U_xV_y - V_xU_y > 0$ in R. Then define T:

$$\begin{cases} \xi = U(x,y) \\ \eta = V(x,y) \end{cases}$$

and it will have all required properties. Carrying through transformation, we get (25.22) with coefficients given by formulas on page 100. This time, however, since U and V are solutions of (25.27),

(25.28) $\quad a \cdot U_x + b \cdot U_y \equiv 0$ and $bV_x + cV_y \equiv 0$ in R_0

and

$$\beta = [aU_x + bU_y]V_x + [bV_x + cV_y]U_y \equiv 0 \text{ in } R.$$

Also since $a \neq 0$ in R, $U_x = -\frac{b}{a} U_y$ and so $\alpha \equiv U_y^2[\frac{b^2}{a} - 2\frac{b^2}{a} + c] \equiv \frac{U_y^2}{a}[ac-b^2] > 0$, since if $U_y = 0$, $U_x = 0$, which is impossible. This α is then $A(\xi,\eta)$. Likewise since $c \neq 0$, $V_y = -\frac{b}{c} V_x$ and we get $\gamma \equiv V_x^2[a - \frac{2b^2}{c} + \frac{b^2}{c}]$

$$= \frac{V_x^2}{c}[ac-b^2] > 0.$$

This γ is $C(\xi,\eta)$.

The transformation used in the above theorem depended essentially upon the fact that $\Delta = ac-b^2 \neq 0$ everywhere in R. Since the same is true for a hyperbolic equation, we have the following, which is sometimes a more useful form than theorem 25.4.

THEOREM 25.6: Under the hypotheses of theorem 25.3, the conclusions given there are still valid when (25.18) is replaced by:

(25.29) $\quad A(\xi,\eta) \cdot \bar{\varphi}_{\xi\xi}(\xi,\eta) - C(\xi,\eta)\cdot\bar{\varphi}_{\eta\eta}(\xi,\eta)+D(\xi,\eta)\cdot\bar{\varphi}_{\xi}(\xi,\eta)+E(\xi,\eta)\cdot \bar{\varphi}_{\eta}(\xi,\eta) + \bar{f}(\xi,\eta,\dots)$

Proof: Set $f_1 = a$, $f_2 = b$, $g_1 = -b$, $g_2 = -c$. Then (25.21) becomes:

(25.30) $\quad a(x,y)\frac{\partial z_1}{\partial x} + b(x,y)\frac{\partial z_1}{\partial y} = 0 \quad ; \quad -b(x,y)\frac{\partial z_2}{\partial x} - c(x,y)\frac{\partial z_2}{\partial y} = 0$

and the solution of these equations gives the required transformation $\xi = U(x,y)$;

$\eta = V(x,y)$. The formulas on page 100 then can be applied to get (25.22). Since (25.28) holds, $\beta = 0$ and, by considering all possibilities, one can easily verify as above that $\alpha > 0$, $\gamma < 0$.

26. The quasi-linear equation in n independent variables is:

$$(26.1)\quad \Sigma\, a_{ik}(x_1,\ldots x_n)\frac{\partial^2 u}{\partial x_i \partial x_k} + f(x_1,\ldots x_n, u, \frac{\partial u}{\partial x_1},\ldots,\frac{\partial u}{\partial x_n}) = 0$$

Without loss of generality, assume $a_{ik} = a_{ki}$.

The quadratic form in n variables

$$Q(\alpha) = \sum_{i,j=1}^{n} A_{ij}\,\alpha_i\alpha_j \qquad ; \quad A_{ij} = a_{ij}(x_1^{\circ},\ldots x_n^{\circ})$$

has the property that if $\alpha_1 = \Sigma\, T_{ij}\beta_j$ where the coefficients form a non-singular matrix T, then $Q(\alpha)$ becomes $\overline{Q}(\beta) = \Sigma\, B_{ij}\beta_i\beta_j$ where $B = ||B_{ij}|| = T'AT$. In particular, there always exists a non-singular transformation T which will reduce $Q(\alpha)$ to the form $\sum_{i=1}^{n} D_i\,\beta_i^2$

where, depending upon the original matrix $A = ||A_{ij}||$,

$$(26.2)\quad \begin{cases} D_i = +1 \text{ for } i=1,2,\ldots,n \quad ;\ n=r=s \\ \text{or}\quad D_i = +1 \text{ for } i=1,\ldots s;\ D_i = -1 \text{ for } i=s+1,\ldots n=r \\ \text{or}\quad D_i = +1 \text{ for } i=1,\ldots s;\ D_i = -1 \text{ for } i=s+1,\ldots,r; \\ \phantom{\text{or}}\quad D_i = 0 \text{ for } i=r+1,\ldots,n. \end{cases}$$

The numbers r and s are invariants under non-singular transformations; they are called the rank and signature, respectively, of T or of $Q(\alpha)$. If $n=r=s$, the form is positive definite and the equation is called elliptic at P°; if $s<n=r$, the equation is called hyperbolic at P°; in particular, if $s=1$, it is called normal hyperbolic; if $s<r<n$, the equation is called parabolic at P°.

If the equation is hyperbolic (or elliptic or parabolic) at every point of a region R, it is said to be hyperbolic (or elliptic or parabolic) in R. Any surface

S: $\qquad x_i = x_i(\mu_1,\ldots\mu_{n-1}) \qquad (i=1,\ldots n)$

which can be written in the form $G(x_1,\ldots,x_n) = \text{const.}$, where $\omega = G(x_1,\ldots,x_n)$ is a solution of:

$$(26.3)\quad Q\left(\frac{\partial\omega}{\partial x_i}\right) \equiv \Sigma\, a_{ij} \frac{\partial\omega}{\partial x_i}\frac{\partial\omega}{\partial x_j} = 0$$

is called a characteristic surface of (26.1). (For a hyperbolic equation, there are n distinct families of characteristic surfaces.) $Q(\alpha)$ is called the characteristic form. When the transformation:

$$\xi_i = t_i(x_1,\ldots,x_n) \qquad (i = 1,..n)$$

where t_i are of class A_1 and

$$J = \left|\frac{\partial t_i}{\partial x_j}\right| \neq 0,$$

is applied to any function $u(x_1,\ldots,x_n)$ of class A^2,

and

$$U(\xi_1,\ldots,\xi_n) = u[t_1^{-1}(\xi_1,\ldots\xi_n),\ldots,t_n^{-1}(\xi_1,..\xi_n)]$$

then

$$u(x_1,\ldots,x_n) = U[t_1(x_1,\cdot,\ x_n),\ldots,t_n(x_1,\ldots,x_n)]$$

The left side of (26.1) becomes

$$(26.4)\quad \Sigma\, b_{i\kappa} \frac{\partial^2 U}{\partial\xi_i\partial\xi_\kappa} + \overline{f}(\xi_1,\ldots,\frac{\partial U}{\partial\xi_n})$$

where

$$b_{ik} = \sum_{l,p} t_{kl}\, a_{lp}\, t_{ip}$$

That is

$$B = J'\,A\,J$$

Now for a particular point $P^0(x_1,\ldots,x_n)$, we can let $A^0 = ||a_{i\kappa}(P_0)||$ and determine T so that $B = T'\,A\,T$ is the diagonal matrix on the previous page. Then the linear transformation:

$$(26.5)\quad \xi_i = \Sigma\, T_{ij}\, x_j$$

will take the left side of (26.1) into (26.2), where at the point $Q^0:(\xi_1^0,..\xi_n^0)$ it becomes $\Sigma\, D_i \frac{\partial^2 U}{\partial\xi_1\partial\xi_k} + \ldots$ One cannot in general find a linear transformation which will do this for every point in the region, but there is one special case where it is possible:

THEOREM 26.1: If A_{ik} is constant, for $i,k=1,\ldots n$, there exists a non-singular matrix $T = ||T_{ij}||$ such that under the linear transformation (26.3), a solution of the equation:

$$(26.6) \quad \Sigma A_{ij} \frac{\partial^2 u}{\partial x_i \partial x_k} + f(x_1,\ldots \qquad) = 0$$

will become a solution of:

$$(26.7) \quad \sum_{i=1}^{n} D_i \frac{\partial^2 U}{\partial \xi_i^2} + \bar{f}(\xi_1,\ldots \qquad) = 0$$

where D_i's are defined in (26.2) for the elliptic, hyperbolic, and parabolic cases, respectively.

When the equation is linear, a further simplification can be made. If

$$(26.8) \quad \Sigma A_{ik} \frac{\partial^2 u}{\partial x_i \partial x_k} + \Sigma A_i \frac{\partial u}{\partial x_i} + A_0 u = h(x_1,\ldots,x_n)$$

where A's are constants, by the last theorem, there exists a transformation (26.4) which reduces the solution of (26.8) to the solution of an equation of the form:

$$(26.9) \quad \Sigma D_i \frac{\partial^2 U}{\partial y_i^2} + \Sigma B_i \frac{\partial U}{\partial y_i} + B_0 U = \bar{h}(y_1,\ldots y_n)$$

the D's being defined in (26.2). Now let

$$V(\xi_1,\ldots,\xi_n) = U(\xi_1,\ldots,\xi_n) e^{\frac{1}{2}\Sigma B_i D_i y_i}$$

If $\Sigma|B_i| = 0$, this will just be U. If $\Sigma|B_i| > 0$, then denoting $\frac{1}{2}\Sigma B_i D_i y_i$ by α

$$U = Ve^{-\alpha}$$

$$U_{y_i} = V_{y_i} e^{-\alpha} - \tfrac{1}{2} B_i D_i V e^{-\alpha}$$

$$U_{y_i y_i} = V_{y_i y_i} e^{-\alpha} - B_i D_i V_{y_i} e^{-\alpha} + \tfrac{1}{4} B_i^2 D_i^2 V e^{-\alpha}$$

Substituting in (26.9) and simplifying, we get:

$$\Sigma D_i \frac{\partial^2 V}{\partial y_i^2} + \Sigma B_i(1 - D_i^2) \frac{\partial V}{\partial y_i} + \frac{V}{4} \Sigma[B_i^2 D_i^3 + B_0] = \bar{h}e^{\alpha}$$

When the equation is elliptic or hyperbolic, $D_i^2 = 1$ for all i and we get,

$$(26.10)\quad \sum_{i=1}^{n} D_i \frac{\partial^2 V}{\partial y_i^2} + (B_0 + \tfrac{1}{4} \Sigma B_i^2 D_i)V = \overline{h}e^{\alpha}$$

In the parabolic case, $D_i^2 = 1$ for 1,...r, and the rest are 0, so we have:

$$(26.11)\quad \sum_{i=1}^{r} D_i \frac{\partial^2 V}{\partial y_i^2} + \sum_{i=r+1}^{n} B_i \frac{\partial V}{\partial y_i} + (B_0 + \tfrac{1}{4}\sum_{i=1}^{r} D_i B_i)V = \overline{h}e^{\alpha}$$

Thus in discussing linear equations with constant coefficients, it is only necessary to consider:

$$(26.12)\quad \sum_{i=1}^{n} \frac{\partial^2 z}{\partial x_i^2} + Kz = f(x_1,\ldots,x_n) \qquad \text{(elliptic)}$$

$$(26.13)\quad \sum_{i=1}^{r} \frac{\partial^2 z}{\partial x_i^2} - \sum_{i=r+1}^{n} \frac{\partial^2 z}{\partial x_i^2} + Kz = f(x_1,\ldots,x_n) \qquad \text{(hyperbolic)}$$

$$(26.14)\quad \sum_{i=1}^{r} \frac{\partial^2 z}{\partial x_i^2} + \sum_{i=r+1}^{n} B_i \frac{\partial z}{\partial y_i} + Kz = f(x_1,\ldots,x_n) \qquad \text{(parabolic)}$$

In a general case the procedure for reduction of (26.8) to (26.10) or (26.11) is given above.

As in Chapter 1, associated with the characteristic equation (26.3), there is a family of ordinary differential equations. If $A(x,p) \equiv \Sigma a(x_1,..x_n)p_i p_j$, they are:

$$(26.15)\quad \frac{dx_i}{ds} = \tfrac{1}{2}\frac{\partial A}{dp_i}, \quad \frac{dp_i}{ds} = -\tfrac{1}{2}\frac{\partial A}{\partial x_i} \qquad (i=1,..n)$$

Any solution of this system which for $s = s_0$ passes through $(x_1^0,..x_n^0, p_1^0,..p_n^0)$ where $A(x^0,p^0) = 0$, is called a bicharacteristic of equation (26.1). The totality of all bicharacteristics through a point forms a surface called the characteristic conoid at that point. Its tangents all form a cone satisfying $A(x^0,p) = 0$. (For constant coefficients, the cone and conoid coincide.) If $q_i = sp_i^0$ and $P_i = sp_i$, the solution of (26.15) can be written: $x_i = \varphi_i(q_1,..q_n;\ x_1^0,...x_n^0)$, $P_i = \psi_i(q_1,..q_n;\ x_1^0,..x_n^0)$ Now set $\Gamma = A(P_1,..P_n;\ x_1,..x_n) = A(q_1,..q_n;\ x_1^0,...x_n^0)$; it is called the geodetic distance from (x) to (x^0). Also, $\Gamma = 0$ is the equation of the characteristic conoid. (Hadamard[1])

B. The Quasi-Linear Hyperbolic Equation

27. The Problem of Goursat

The first basic problem for the hyperbolic equation:

$$(23.1)\quad a(x,y)\frac{\partial^2 z}{\partial x^2} + 2b(x,y)\frac{\partial^2 z}{\partial x\partial y} + c(x,y)\frac{\partial^2 z}{\partial y^2} + f\left(x,y,z,\frac{\partial z}{\partial x},\frac{\partial z}{\partial y}\right) = 0 \qquad (ac < b^2)$$

may be stated as follows:

Problem G: (a) Let $f(x,y,z,p,q)$ be defined for $(x,y)\ \varepsilon R_0$ and $(z,p,q)\ \varepsilon V$. Let $a(x,y)$, $b(x,y)$, $c(x,y)$ be functions of class A^2 with $ac < b^2$ everywhere in R_0. (Perhaps other conditions.)

(b) Let (x_0,y_0) be a point in R_0, and let characteristics of (23.1) through (x_0,y_0) be:

$$\Gamma_1: \begin{cases} x = x(\mu) \\ y = y(\mu) \end{cases} \qquad \Gamma_2: \begin{cases} x = \bar{x}(\nu) \\ y = \bar{y}(\nu) \end{cases}$$

where these are defined as solutions of 25.1 (see Corollary 1, p. 96) for $\mu \varepsilon M_1$: $\mu_1 < \mu < \mu_2$ and $\nu \varepsilon M_2$: $\nu_1 < \nu < \nu_2$. (Let the range of

$x(\mu)$ be S_1, of $y(\mu)$ be T_1; taking $S = S_1 \cdot S_2$ and

$\bar{x}(\nu)$ be S_2 $\quad \bar{y}(\nu)$ be T_2

$T = T_1 \cdot T_2$, we have Γ_1 and Γ_2 in $S \times T$, provided $(x_0, y_0) \varepsilon S \times T$.)

(c) Let $\sigma(\mu)$ be defined for $\mu \varepsilon M_1$ and $\tau(\nu)$ be defined for $\nu \varepsilon M_2$. Let $\sigma(\mu_0) = \tau(\nu_0)$ where μ_0 is the value of the parameter corresponding to (x_0, y_0) on Γ_1 and ν_0 is the value for (x_0, y_0) on Γ_2.

Then we wish to establish the existence of a function $\varphi(x,y)$ such that:

(α) $\varphi(x,y)$ is of class A^2 in $R \leq R_0$

(β) $z = \varphi(x,y)$ is a solution of (23.1) in R_0

(γ) $\varphi(x(\mu), y(\mu)) = \sigma(\mu)$ for $\mu \varepsilon \bar{M}_1 \leq M_1$ and $\varphi(\bar{x}(\nu), \bar{y}(\nu)) = \tau(\nu)$ for $\nu \varepsilon M_2 \leq M_2$.

We shall solve this problem first for the equation

$$(27.1) \quad \frac{\partial^2 z}{\partial x \partial y} - f(x,y,z,\frac{\partial z}{\partial x},\frac{\partial z}{\partial y}) = 0.$$

For this equation, $A = C = 0$, $B = 1$, so it is hyperbolic in R_0 (f is defined for $(x,y) \varepsilon R_0$) and the characteristic curves (see theorem 25.2) through (x_0, y_0) become:

$$(27.2) \quad x = x_0 \quad \text{and} \quad y = y_0$$

THEOREM 27.1: (a_0) Let $f(x,y,z,p,q)$ be continuous for $(x,y) \varepsilon R_0$ and all z,p,q. In any closed bounded subrectangle of R_0, $R = S \times T$, where S is: $x_1 \leq x \leq x_2$ and T is: $y_1 \leq y \leq y_2$, let f satisfy a Lipschitz condition:

$$(27.3) \quad |f(x,y,z,p,q) - f(x,y,\bar{z},\bar{p},\bar{q})| \leq M[|\bar{z}-z| + |\bar{p}-p| + |\bar{q}-q|]$$

(b) Let (x_0, y_0) be a point in R_0

(c) Let $\sigma(x)$ be of class A^1 in $S_1 : \alpha<x<\beta$; let $\tau(y)$ be of class A^1 in $T_1 : \gamma<y<\delta$, where $R_1 = S_1 \times T_1 \leq R_0$ and let $(x_0, y_0) \varepsilon R_1$. Let $\sigma(x_0) = \tau(y_0)$. Then there exists a unique function $\varphi(x,y)$ such that

(α) $\varphi(x,y)$ is defined and of class A^2 in R_1

(β) $z = \varphi(x,y)$ is a solution of (27.1) in R_1

(γ) $\varphi(x_0, y) \equiv \tau(y)$ for $y \varepsilon T_1$ and $\varphi(x, y_0) \equiv \sigma(x)$ for $x \varepsilon S_1$.

Proof: By Picard's method of successive approximation, define:

$$\varphi^0(x,y) = \sigma(x) + \tau(y) - \sigma(x_0)$$

$$\varphi^n(x,y) = \varphi^0(x,y) + \int_{x_0}^{x}\int_{y_0}^{y} f[t,s,\varphi^{n-1}(t,s),\varphi_x^{n-1}(t,s),\varphi_y^{n-1}(t,s)]\,dt\,ds$$

$$(n=1,2,3,\ldots,)$$

Then

$$\varphi_x^n(x,y) = \int_{y_0}^{y} f[x,s,\ldots,\varphi_y^{n-1}(x,s)]\,ds$$

$$\varphi_y^n(x,y) = \int_{x_0}^{x} f[t,y,\ldots,\varphi_y^{n-1}(t,y)]\,dt$$

In every closed subrectangle $R < R_1$, all three sequences converge uniformly:

$$\left|\varphi^{n+1} - \varphi^n\right| \leq \frac{A}{3M}\,\frac{(3KM)^{n+1}}{(n+1)!}\left(|x-x_0| + |y-y_0|\right)^{n+1} \quad \text{in } R = S \times T$$

where $|f| \leq A$ and $K = \max\left(1, \frac{x_2-x_1}{2}, \frac{y_2-y_1}{2}\right)$.

Let $\varphi(x,y) = \lim_{n\to\infty} \varphi^n(x,y)$. It will satisfy α, β, γ.

Proof found in many places. See Kamke [1]; Tamarkin and Feller [1].

THEOREM 27.2

(a) Let $f(x,y,z,p,q)$ be continuous for $(x,y)\varepsilon R_0$ and z,p,q arbitrary.

In any closed bounded subregion of R_0, let f satisfy a Lipschitz condition (27.3).

(a_1) Let a(x,y), b(x,y), c(x,y) be of class A^2 in R_0 and let $ac < b^2$ in R_0.

(b) Let the characteristics Γ_1 and Γ_2 through a point (x_0, y_0) of R_0 be such that for $\mu \varepsilon M_1$ and $\nu \varepsilon M_2$ Γ_1 satisfies (25.7) Γ_2 satisfies (25.9), if $\lambda_1 \neq 0$. [Use (25.13) and (25.11) if $\lambda_1 = 0$, $\lambda_2 \neq 0$]

(c) Let $\sigma(\mu)$ be of class A^1 for $\mu \varepsilon M_1$, let $\tau(\nu)$ be of class A^1 for $\nu \varepsilon M_2$, and let $\sigma(\mu_0) = \tau(\nu_0)$ where $\begin{cases}\mu_0\\ \nu_0\end{cases}$ is the value of the parameter corresponding to (x_0, y_0) on $\begin{cases}\Gamma_1\\ \Gamma_2\end{cases}$. Then there exists a function $\varphi(x,y)$ such that

(α) $\varphi(x,y)$ is of class A^2 in some $R \leq R_0$

(β) $z = \varphi(x,y)$ is a solution of (23.1) in R

(δ) $\varphi(x_0(\mu), y_0(\mu)) = \sigma(\mu)$ and $\varphi(\bar{x}(\nu), \bar{y}(\nu)) = \tau(\nu)$

Proof: By hypothesis there is a certain neighborhood N of (x_0, y_0) where (25.7) and (25.9) have unique solutions:

$$\Gamma_1 \begin{cases} x = x(\mu) \\ y = y(\mu) \end{cases} \quad \text{and} \quad \Gamma_2 \begin{cases} x = \bar{x}(\nu) \\ y = \bar{y}(\nu). \end{cases}$$

Now define U(x,y) and V(x,y) as in theorem 25.3, page 98, and perform the transformation

$$T: \begin{cases} \xi = U(x,y) \\ \eta = V(x,y) \end{cases}$$

This will take (x_0, y_0) into $(\xi_0 = U(x_0, y_0),\ \eta_0 = V(x_0, y_0))$ and the characteristics Γ_1 into L_1: $\xi = \xi_0$ and Γ_2 into L_2: $\eta = \eta_0$. That is, Γ_1 will go into a segment of $\xi = \xi_0$, corresponding to $\eta_1 < \eta < \eta_2$ and Γ_2 will go into a segment of $\eta = \eta_0$, corresponding to $\xi_1<\xi<\xi_2$ where $\eta_1 = \lim_{\mu \to \mu_1^+} V[x(\mu), y(\mu)]$, $\eta_2 = \lim_{\mu \to \mu_2^-} V[x(\mu), y(\mu)]$ and $\xi_1 = \lim_{\nu \to \nu_1^+} U[\bar{x}(\nu),\bar{y}(\nu)]$ and where $\xi_2 = \lim_{\nu \to \nu_2^-} U[\bar{x}(\nu),\bar{y}(\nu)]$. Since the points of Γ_1 and Γ_2 were in R_0, the points of L_1 and L_2 are in $\bar{R}_0$, so the rectangle $\bar{R}_1 = S_1 \times T_1 \leq \bar{R}_0$ where S_1: $\xi_1<\xi<\xi_2$ and T_1: $\eta_1<\eta<\eta_2$.

One can readily show as in the corollary, page 101, that if $h(\mu) = V[x(\mu),y(\mu)]$, $h'(\mu) \neq 0$ in M_1, that is, on Γ_1 the rate of change of V is not zero. Hence $\eta = V[x(\mu),y(\mu)] \equiv h(\mu)$ has a unique inverse $\mu = S_1(\eta)$ at least for $\eta \varepsilon T_1$ and as μ varies so that $x = x(\mu)$, $y = y(\mu)$ describes Γ_1, η varies so that $\xi = \xi_0$, $\eta = \eta$ describes L_1, where $\eta = h(\mu)$ or $\mu = S_1(\eta)$. Hence if we wish to find a solution such that $\varphi(x(\mu),y(\mu)) = \sigma(\mu)$, on Γ_1, $\bar{\varphi}(\xi\ \eta) \equiv \varphi[X(\xi,\eta),Y(\xi,\eta)]$ must have the property that on L_1, $\bar{\varphi}(\xi_0,\eta) = \sigma(\mu)$ where $\mu = S_1(\eta)$. So define $\bar{\tau}(\eta) = \sigma(S_1(\eta))$. Likewise $k(\nu) = U[x(\mu),y(\mu)]$ has $k'(\nu) \neq 0$ in M_2, and hence $\xi = U[x(\nu),y(\nu)] \equiv k(\nu)$ has a unique inverse $\nu = S_2(\xi)$ for $\xi \varepsilon S_1$, and as ν varies and describes Γ_2: $x = \bar{x}(\nu)$, $y = \bar{y}(\nu)$, ξ varies so that $\xi = \xi$, $\eta = \eta_0$ describes L_2, where $\xi = k(\mu)$. Thus if we wish to have $\varphi(\bar{x}(\nu), \bar{y}(\nu)) = \tau(\nu)$ on Γ_2, then defining $\bar{\sigma}(\xi) = \tau(S_2(\xi))$, it necessarily follows that $\bar{\varphi}(\xi,\eta_0) = \bar{\sigma}(\xi)$ on L_2.

Under T, by theorem 25.3 we need only be concerned with the solution of

$$(27.6) \quad \frac{\partial^2\omega}{\partial\xi\partial\eta} + F(\xi,\eta,\omega,\bar{p},\bar{q}) = 0$$

where $F(\xi,\eta,\omega,\bar{p},\bar{q}) \equiv \frac{1}{2B(\xi,\eta)} [D(\xi,\eta) \cdot \bar{p} + E(\xi,\eta) \cdot \bar{q} + f(X(\xi,\eta),Y(\xi,\eta),\omega,$

$\bar{p}U_x + \bar{q}V_x, \bar{p}U_y + \bar{q}V_y$) (D, E, and B are defined in theorem 25.3). Now, in any bounded closed region of $\bar{R}$:

$$|F(\xi,\eta,\omega_1,\bar{p}_1,\bar{q}_1) - F(\xi,\eta,\omega_2,\bar{p}_2,\bar{q}_2)| \leq M_1[|\bar{p}_1-\bar{p}_2|] + M_2[|\bar{q}_1-\bar{q}_2|]$$

$$+ |f(\xi,\eta,\omega_1,\bar{p}_1,\bar{q}_1) - f(\xi,\eta,\omega_2,\bar{p}_2,\bar{q}_2)| \leq \bar{M}[|\bar{p}_1-\bar{p}_2| + |\bar{q}_1-\bar{q}_2|$$

$+ |\omega_1-\omega_2|]$, by hypothesis on f. We then apply theorem 27.1, page 110 to find a solution $\omega = \bar{\varphi}(\xi,\eta)$ of 27.6 such that $\bar{\varphi}(\xi_0,\eta) = \bar{\tau}(\eta)$ and $\bar{\varphi}(\xi,\eta_0) = \bar{\sigma}(\xi)$ in $\bar{R}_1 = \bar{S}_1 \times \bar{T}_1$. Such a solution exists and is unique. So

$$\bar{\varphi}_{\xi\eta}(\xi,\eta) + F(\xi,\eta,\varphi,\varphi_\xi,\varphi_\eta) \equiv 0.$$

By theorem 25.3, then, if we set $\varphi(x,y) \equiv \bar{\varphi}[U(x,y),V(x,y)]$, $z = \varphi(x,y)$ is a solution of (23.1) in R_1, the inverse map of $\bar{R}_1$, and $\varphi(x(\mu),y(\mu)) = \bar{\tau}(\eta) = \sigma(S_1(\eta)) = \sigma(\mu)$ and likewise on Γ_2.

COROLLARY 1: If (a_0), (b), (c) hold where (23.1) is of the form:

$$(23.4)\quad A\frac{\partial^2 z}{\partial y^2} + 2B\frac{\partial^2 z}{\partial x\partial y} + C\frac{\partial^2 z}{\partial y^2} + f(x,y,z,\frac{\partial z}{\partial x},\frac{\partial z}{\partial y}) = 0$$

where A, B, C are constants and $AC < B^2$, there exists a function $\varphi(x,y)$ satisfying (α), (β), (γ). It is $\varphi(x,y) \equiv \bar{\varphi}(x + \frac{A}{\lambda_1}y,\ y + \frac{C}{\lambda_1}x)$

where $\omega = \bar{\varphi}(\xi,\eta)$ is the solution of problem G for (27.6) under conditions

$$\varphi(\xi_0,\eta) = \bar{\tau}(\eta) \text{ and } \varphi(\xi,\eta_0) = \bar{\sigma}(\eta)$$

The transformation is

$$T: \begin{cases} \xi = x + \frac{A}{\lambda_1}y \\ \eta = y + \frac{C}{\lambda_1}x \end{cases}$$

with Jacobian $J = \frac{2\sqrt{B^2-AC}}{\lambda_1}$; $\lambda_1 = -B + \sqrt{B^2-AC}$ (assuming $B \leq 0$); (ξ_0,η_0)

is a point corresponding to (x_0, y_0) under T

$$\bar{\tau}(\eta) = \sigma\left(\frac{\eta-\eta_0}{J}\right) \text{ and } \bar{\sigma}(\xi) = \tau\left(\frac{\xi-\xi_0}{J}\right).$$

The solution exists in a parallelogram P contained in R with center (x_0, y_0) and sides parallel to Γ_1 and Γ_2 respectively. In (27.6)

$$F(\xi,\eta,\omega,\bar{p},\bar{q}) \equiv \frac{-1}{2J\sqrt{B^2-AC}} f\left[\frac{\lambda_1\xi-A\eta}{2\sqrt{B^2-AC}}, \frac{\lambda_1\eta-C\xi}{2\sqrt{B^2-AC}}, \omega, \bar{p} + \frac{C}{\lambda_1}\bar{q}, \bar{p} + \frac{A}{\lambda_1}\bar{q}\right].$$

Proof: Corollary 1, page 96, gave formulas for Γ_1 and Γ_2; writing them

$$x + \frac{A}{\lambda_1} y = \xi_0 \text{ and } y + \frac{C}{\lambda_1} x = \eta_0 \text{ where } \xi_0 = x_0 + \frac{A}{\lambda_1} y_0,$$

$$\eta_0 = y_0 + \frac{C}{\lambda_1} x_0$$

it is easy to see that

$$U(x,y) \equiv x + \frac{A}{\lambda_1} y$$

$$V(x,y) \equiv y + \frac{C}{\lambda_1} x$$

give the transformations of the theorem, taking Γ_1 into $\xi = \xi_0$ and Γ_2 into $\eta = \eta_0$. Along Γ_1, $V(x(\mu),y(\mu)) = \eta_0 + \varphi J$, so inverse of $\eta = \eta_0 + J\mu$ is $\mu = \frac{\eta-\eta_0}{J}$ and on Γ_2, $U(x(\bar{v}),y(\bar{v})) = \xi_0 + vJ$, so inverse of $\xi = \xi_0 + Jv$ is

$v = \frac{\xi-\xi_0}{J}$. Defining $p = \bar{p} + \frac{C}{\lambda_1}\bar{q}$, $q = \bar{q} + \frac{A}{\lambda_1}\bar{p}$, we have correct $\bar{p}$ and $\bar{q}$.

Note that here inverse $X(\xi,\eta) = \frac{1}{J}\left[\xi - \frac{A}{\lambda}\eta\right] = \frac{\lambda_1\xi-A\eta}{2\sqrt{B^2-AC}}$ and $Y(\xi,\eta) = \frac{1}{J}\left[\eta-\frac{C}{\lambda}\xi\right]$

$= \frac{\lambda_1\eta\cdot C\xi}{2\sqrt{B^2-AC}}$ Hence $\bar{q} = \frac{\lambda_1 q-Ap}{2\sqrt{B^2-AC}}$ and $\bar{p} = \frac{\lambda_1 p-Cq}{2\sqrt{B^2-AC}}$.

Schauder[1] has used his fixed point theorem to establish existence theorems of various kinds in several different types of differential equations; it is an interesting and important method, although it does not appear too practical. We give the following theorem, which concerns a variation of problem G to which the general one can be reduced in many instances:

THEOREM 27.3: Let f be defined for $|x| \leq d$, $|y| \leq d$, $|z| \leq d$, $|p| \leq d$, $|q| \leq d$ and satisfy a Holder condition with constant C and exponent λ $(0 < \lambda < 1)$ with respect to x,y,z and a Lipschitz condition with respect to p and q:

$$|f(x,y,z,p,q) - f(\bar{x},\bar{y},\bar{z},\bar{p},\bar{q})| \leq C[|x-\bar{x}|^{\lambda_1}+|y-\bar{y}|^{\lambda_2}+|z-\bar{z}|^{\lambda_3}+|p-\bar{p}|+|q-\bar{q}|]$$

Then there exists a unique solution of (27.1) such that

$$\varphi(x,0) \equiv \varphi(0,y) \equiv 0 \qquad (f(0,\ldots,0) = 0)$$

Proof: Schauder[1]. The fixed point theorem itself states that if H is an arbitrary closed set in a vector field and F(e) is a continuous operator which takes every element of H into an element of H, then there exists an e_0 such that $e_0 = F(e_0)$. In this instance, Q is the set of all functions z(x,y) which are of class A^2 for $|x| \leq d, |y| \leq d$ and have $z_{xy}(0,0) = z(x,0) = z(0,y)=0$. It is shown that this is a vector field and that the operator U(z) = $\int_0^x\int_0^y f(x,y,z,z_x,z_y)dydx$ makes correspond to every function of Q a function of Q. Every subset of Q in which z_{xy} satisfies a Holder condition with constant M and exponent μ forms a closed, compact, convex set $H_{M\mu}$; choose M and μ so that in a subset of R,U[z] ε $H_{M\mu}$. Hence there is a fixed point in H. That is, there is a function z_0 of the desired type such that $z_0(x,y) = U(z_0(x,y))$, which is equivalent to saying z_0 satisfies (27.1).

Sato[1] used the same method to show that theorem 27.1 is still valid when condition (a_0) is replaced by the following: Let f be continuous in a domain D and let there exist two functions $\underline{\omega}$ and $\overline{\omega}$ in D with $\underline{\omega}_{xy} \leq f(x,y,z,z_x,z_y)$

$\leq \overline{\omega}_{xy}$ whenever $\underline{\omega} \leq z \leq \overline{\omega}$, $\underline{\omega}_x \leq z_x \leq \overline{\omega}_x$, $\underline{\omega}_y \leq z_y \leq \overline{\omega}_y$.

28, The second basic problem for quasi-linear equations is the problem of Cauchy.

Problem C: (a) Let $f(x,y,z,p,q)$ be defined for $(x,y) \varepsilon R_0$ and $(x,p,q) \varepsilon V$. Let $a(x,y)$, $b(x,y)$, $c(x,y)$ be functions of class A^2 with $ac < b^2$ everywhere in R_0.

($\overline{b}$) Let C: $\begin{matrix} x = x_0(t) \\ y = y_0(t) \end{matrix}$ be a curve of class A^1 for $t \varepsilon T$: $t_1 < t < t_2$ and such that $(x(t), y(t)) \varepsilon R_0$ for $t \varepsilon T$. Let C not be tangent to a characteristic of (23.1) at any point.

(d) Let $z_0(t)$, $p_0(t)$, $q_0(t)$ be defined for $t \varepsilon T$, where $z_0(t)$ is of class A^1, $p_0(t)$ and $q_0(t)$ are of class A^0 and

(28.1) $z_0'(t) = p_0(t) \cdot x_0'(t) + q_0(t) \cdot y_0'(t)$

Then we wish to establish the existence of a function $\varphi(x,y)$ such that

(α) $\varphi(x,y)$ is of class A^2 in $R \leq R_0$

(β) $z = \varphi(x,y)$ is a solution of (23.1) in R_0

(δ) For $t \varepsilon T_1 \leq T$,

(28.2) $\varphi(x_0(t), y_0(t)) \equiv z_0(t)$; $\varphi_x(x_0(t), y_0(t)) \equiv p_0(t)$;

$\varphi_y(x_0(t), y_0(t)) \equiv q_0(t)$

(The condition that C not be tangent to a characteristic insures uniqueness.)

As in Chapter I, when $x_0'(t) \neq 0$, we can write C in the form $y = y_0(x)$ and if $y_0'(t) \neq 0$, we can write C in the form $x = x_0(y)$. This leads to variations of problem C.

Problem C_0: (a) As above

($\overline{\overline{b}}$) Let $x = \overline{x}_0(y)$ be a curve C, where $\overline{x}_0(y)$ is of class A^1 for $y \varepsilon T$: $y_1 < y < y_2$.

($\overline{\overline{d}}$) Let $z_0(y)$, $p_0(y)$, $q_0(y)$ be given as in (d) with

(28.3) $\overline{z}_0(y) = \overline{p}(y) \cdot \overline{x}_0(y) + q_0(y)$

Then we wish to establish the existence of a function $\varphi(x,y)$ satisfying (α), (β) and $(\bar{\delta})$ For $y\varphi T \leq T_1$,

$$(28.4) \quad \varphi(\bar{x}_0(y),y) = \bar{z}_0(y), \ \varphi_x(x_0(y),y) \equiv p_0(y), \ \varphi_y(\bar{x}_0(y),y) \equiv q_0(y)$$

However, in (28.3), $q_0(y)$ is completely determined, so it may be omitted from given conditions and last equation of (28.4) omitted. In other words:

Problem C_1: (a) As above; $(\bar{\bar{b}})$ as above;

(d_1): Let $z_0(y)$ and $p_0(y)$ be given.

Then we wish to establish existence of a function $\varphi(x,y)$ satisfying (α) (β) and:

$$(28.5) \quad \begin{cases} \varphi(x_0(y),y) = \bar{z}_0(y) \\ \varphi_x(x_0(y),y) = p_0(y) \end{cases}$$

A particular case of C_1 is the so-called *initial value problem:*

Problem I: (a) As above

(d_0) Let $\bar{z}_0(y)$ and $\bar{p}_0(y)$ be defined for $y\varepsilon T$; and let x_0 be a given number such that $(x_0,y)\varepsilon R$ for all $y\varepsilon T$. Then we wish to establish the existence of a function $\varphi(x,y)$ such that (α), (β) and:

(δ_0) $\varphi(x_0,y) = \bar{z}_0(y)$; $\varphi_x(x_0,y) = \bar{p}_0(y)$ for all $y\varepsilon T$.

[When $x_0 = 0$, we call the problem I_0]

As in the case of Problem G, we shall first establish solutions of these problems for the canonical form:

$$(27.1) \quad \frac{\partial^2 z}{\partial x \partial y} = f(x,y,z,\frac{\partial z}{\partial y},\frac{\partial^2 z}{\partial y^2})$$

and then in general. When the equation has $b \equiv 0$ (which is a form to which the general equation can be reduced by theorem 25.4), one can obtain formulas for solution directly.

Sometimes C is given by specifying $p_0(t)$ and $q_0(t)$ and simply requiring that $\varphi(x_0, y_0) = z_0$, for some point. For if $\varphi_x(x_0(t), y_0(t)) = p_0(t)$ and $\varphi_y(x(t), y(t)) = q_0(t)$, then the strip equation will define

$$\varphi(x_0(t), y_0(t)) \equiv \int_{t_0}^{t} [p_0 \cdot x_0' + q_0 \cdot y_0'] \, d\bar{t}$$

So knowing a single point, $z_0(t)$ is completely determined.

The first theorem has to do with a special kind of Cauchy problem - we may call it problem $\bar{C}$.

THEOREM 28.1

(b) Let

$$x = x_0(\mu)$$
$$y = y_0(\mu)$$

be a curve C, where $x_0(\mu)$ and $y_0(\mu)$ are of class A^1 in M: $\mu_1 < \mu < \mu_2$ and $x_0' \neq 0$, $y_0'(\mu) \neq 0$ in M. Set $\alpha = \lim_{\mu \to \mu_1^+} x_0(\mu)$, $\beta = \lim_{\mu \to \mu_2^-} t_0(\mu)$, if $x_0' > 0$, (otherwise reverse definition), and $\gamma = \lim_{\mu \to \mu_1^+} y_0(\mu)$, $\delta = \lim_{\mu \to \mu_2^-} y_o(\mu)$ if $y_o' > 0$, (if $y_o' < 0$ reverse).

(d) Let

$\sigma(x)$ be a function of class A^1 in S: $\alpha < x < \beta$

$\tau(y)$ be a function of class A^1 in T: $\gamma < y < \delta$

(a_0) Let $f(x,y,z,p,q)$ be continuous in $U = R \times V$, where V is an arbitrary region, and $R = S \times T$; let f be bounded in $V_0 = R_0 + V$, where R is a closed bounded subrectangle $R_0 = S_0 \times T_0 \leq R$, and let satisfy a Lipschitz condition (27.3) in R_0. Then there exists a unique function $\varphi(x,y)$ such that

(α) $\varphi(x,y)$ is of class A^2 in R

(β) $z = \varphi(x,y)$ is a solution of (27.1) in R

(γ) For $\mu \varepsilon M$, $\varphi(x_0(\mu), y_0(\mu)) = \sigma(x_0(\mu)) + \tau(y_0(\mu))$.

Proof: As on page 8, let $\mu = s_1(x)$ be the inverse function of $x = x_0(\mu)$; it will be of class A^1 in S; likewise $\mu = s_2(y)$, the inverse function of $y = y_0(\mu)$ will be of class A^1 in T; consider, for a given x_1, y_1 in R, the closed region bounded by C , the straight line $x = x_1$ and the line $y = y_1$; call it $R_{x_1 y_1}$. Then define approximating functions as follows:

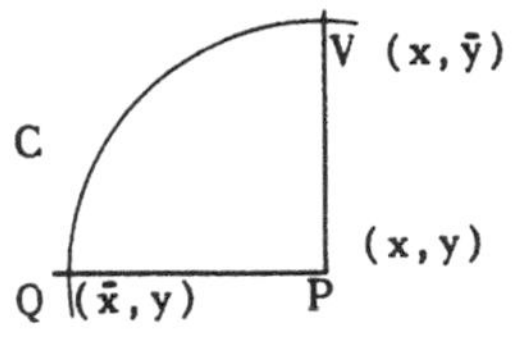

$$\varphi^0(x,y) = \sigma(x) + \tau(y)$$

$$\varphi^n(x,y) = \varphi^0(x,y) \pm \iint_{R_{xy}} f(t,s,\varphi^{n-1}(t,s),\varphi_x^{n-1}(t,s),\varphi_y^{n-1}(t,s))\,dsdt$$

$(n=1,2,3\ldots)$

(use + sign if P lies above curve, - sign if it lies below)

Then $\varphi^n(x,y)$ is of class A^2 and

$$\varphi_x^n(x,y) = \int_{\bar{y}}^{y} f(x,s,\ldots)\,ds \quad \text{where } \bar{y} = y_0[s_1(x)]$$

$$\varphi_y^n(x,y) = \int_{\bar{x}}^{x} f(t,y,\ldots)\,dt \quad \text{where } \bar{x} = x_0[s_2(y)]$$

As in previous theorem, $\overset{n}{\varphi}(x,y)$ converges uniformly in R_0 (same bound on $|\overset{n}{\varphi}-\overset{n-1}{\varphi}|$), and if we set $\varphi(x,y) = \lim_{n\to\infty} \overset{n}{\varphi}(x,y)$, this will be desired result.

COROLLARY: (d_0) Let $x = \bar{x}_0(y)$, where $\bar{x}_0(y)$ is of class A^1 in T: $\gamma<y<\delta$ and $\bar{x}_0'(y) \neq 0$, be a curve C.

(a_0) Let $\bar{\tau}(y)$ be a function of class A^1 in T.

(c) Let f be defined as in theorem.

Then there exists a unique function $\varphi(x,y)$ such that:

(α) $\varphi(x,y)$ is of class A^2 in R.

(β) $z = \varphi(x,y)$ is a solution of (27.1) in R.

(γ) For all y in T, $\varphi(x_0(y),y) = \bar{\tau}(y)$.

Proof: As on page 8, the given C can be represented as

$$\begin{cases} x = x_0(\mu) \\ y = y_0(\mu) \end{cases}$$

if we take $\tau_0(\mu) = \bar{\tau}(\mu), x_0(\mu) = \bar{x}_0(\mu)$, $y_0(\mu) = \mu$. Then apply theorem.

Conversely, given a solution to the problem of the corollary, we can get the solution to $\overline{C}$ from it, by setting $\bar{x}_0(y) = x_0[s_2(y)]$ and $\bar{\tau}(y) = \sigma(x_0(s_2(y))) + \tau(y)$.

THEOREM 28.2. (d) Let $\begin{cases} x = x_0(\mu) \\ y = y_0(\mu) \end{cases}$

be a curve as defined in theorem 28.1.

($\bar{a}$) Let $z_0(\mu), p_0(\mu), q_0(\mu)$ be defined on M and $z_0'(\mu), p_0(\mu)$,

$q_0(\mu)$ be continuous there; suppose

$$(25.2) \quad z_0'(\mu) = p_0(\mu) \cdot x_0'(\mu) + q_0(\mu) \cdot y_0'(\mu)$$

($\bar{c}$) Let $f(x,y,z,p,q)$ be continuous in a region U which contains $(x_0(\mu),\ldots,q_0(\mu))$, bounded wherever $(x,y)\varepsilon R_0$, and satisfying (27.3).

Then there exists a unique function $\varphi(x,y)$ such that

(α) $\varphi(x,y)$ is of class A^2 in R

(β) $\varphi(x,y)$ is a solution of (27.1) in R

($\bar{\gamma}$) For $\mu\varepsilon M$, $\varphi(x_0(\mu),y_0(\mu)) = z_0(\mu)$; $\varphi_x(x_0(\mu),y_0(\mu)) = p_0(\mu)$;

$\varphi_y(x_0(\mu),y_0(\mu)) = q_0(\mu)$

[Knowing the answer to problem $\bar{C}$, we can apply same formula here by taking

$$\sigma(x) = \int_{x_0}^{x} p(s_1(t))\, dt + \tfrac{1}{2}z_0, \quad \tau(y) = \int_{y_0}^{y} q(s_2(t))\, dt + \tfrac{1}{2}z_0$$

where μ_0 is some fixed value in M, $x_0 = x_0(\mu_0)$, $y_0 = y_0(\mu_0)$, $z_0 = z_0(\mu_0)$ and $\mu = s_1(x)$ is the inverse of $x = x_0(\mu)$, $\mu = s_2(y)$ is the inverse of $y = y_0(\mu)$]

Proof: Defining $\sigma(x)$ and $\tau(y)$ as above, the hypotheses of theorem 28.1 are satisfied and we can get a unique $\varphi(x,y)$ which satisfies α, β, γ. Hence, for all μ on M, $\varphi(x_0(\mu),y_0(\mu)) = \sigma(x_0(\mu)) + \tau(y_0(\mu)) = \int_{x_0}^{x} p(s_1(t))\, dt$ $+ \int_{y_0}^{y} q(s_2(t))\, dt + z_0 = \int_{\mu_0}^{\mu} p(\bar{\mu}) \cdot x_0'(\bar{\mu})\, d\bar{\mu} + \int_{\mu_0}^{\mu} q(\bar{\mu}) \cdot x_0^{1'}\, d\bar{\mu} + z_0$ $= z_0(\mu)$ by (25.2) and $\varphi_x(x_0(\mu),y_0(\mu)) = \sigma'(x_0(\mu)) = p(s_1(x_0(\mu))) = p(\mu)$; $\varphi_y(x_0(\mu),y_0(\mu)) = q(\mu)$, so ($\bar{\gamma}$) is satisfied.

COROLLARY 1: (d_0) Let $x = \bar{x}_0(y)$ where $\bar{x}_0(y)$ is of class A^1 in T: $\gamma<y<\delta$ and $\bar{x}_0'(y) \neq 0$, be a curve C.

(a_0) Let $z_0(y)$ be a function of class A^1 in T, and $p_0(y)$ be continuous on T.

$(\bar{c})$ Let f be defined as in the above theorem.

Then there exists a unique function $\varphi(x,y)$ such that

(α) $\varphi(x,y)$ is of class A^2 in R

(β) $\varphi(x,y)$ is a solution of (27.1) in R

$(\bar{\gamma})$ For $y \varepsilon T$, $\varphi(\bar{x}_0(y),y) = z_0(y)$ and $\varphi_x(\bar{x}_0(y),y) = p_0(y)$

Proof: Set $q_0(y) = z_0'(y) - p_0(y) \cdot x_0'(y)$. Let $x_0(\mu), y_0(\mu) = \mu$ and apply theorem.

We now return to the general quasi-linear equation (23.1)

THEOREM 28.3: (a_0) Let $f(x,y,z,p,q)$ be continuous for $(x,y) \varepsilon R_0$ and (z,p,q) arbitrary. In every bounded closed subregion of R_0, let f satisfy a Lipschitz condition (27.3) and be bounded.

(a_1) Let $a(x,y)$, $b(x,y)$, $c(x,y)$ be of class A^2 in R_0 and let $ac < b^2$ in R_0.

$(\bar{b})$ Let C: $\begin{cases} x = x_0(t) \\ y = y_0(t) \end{cases}$ be a curve of class A^1 for $t \varepsilon T$:

$t_1 < t < t_2$ and such that $(x(t),y(t)) \varepsilon R_0$ for $t \varepsilon T$. Suppose C is not tangent to a characteristic of (23.1) at any point.

(d) Let $z_0(t)$, $p_0(t)$, $q_0(t)$ be defined for $t \varepsilon T$, where $z_0(t)$ is of class A^1, $p_0(t)$ and $q_0(t)$ are of class A^0 and

(28.1) $z_0'(t) = p_0(t) \cdot x_0'(t) + q_0(t) \cdot y_0'(t)$

Then there exists a function $\varphi(x,y)$ such that

(α) $\varphi(x,y)$ is of class A^2 in R_0

(β) $z = \varphi(x,y)$ is a solution of (23.1) in R_0

(δ) For $t \varepsilon T_1 \leq T_0$

(28.2) $\varphi(x_0(t), y_0(t)) \equiv z_0(t);\ \varphi_x \equiv p_0(t);\ \varphi_y \equiv q_0(t)$

Proof: Defining U(x,y) and V(x,y) as in theorem 25.2. let

$$\begin{cases} \xi = U(x,y) \\ \eta = V(x,y) \end{cases}$$

then the curve $C \to \bar{C}$: $\begin{cases} \xi = \xi_0\ (t) \equiv U[x_0(t), y_0(t)] \\ \eta = \eta_0\ (t) \equiv V[x_0(t), y_0(t)] \end{cases}$

Also let $\begin{cases} \bar{p}_0(t) = p(t) \cdot X_\xi + q \cdot Y_\xi \\ \bar{q}\ (t) = p \cdot X_\eta + q \cdot Y_\eta \end{cases}$ where $\begin{cases} X = X[\xi_0(t), \eta_0(t)]. \\ Y = Y[\xi_0(t), \eta_0(t)] \end{cases}$

Now the condition that C not be tangent to a characteristic means that

$\begin{cases} \xi_0'(t) \neq 0, \\ \eta_0'(t) \neq 0 \end{cases}$ since characteristics of transformed equations are $\begin{cases} \xi = \xi_0. \\ \eta = \eta_0 \end{cases}$

$[\xi_0'(t) = U_x \cdot x_0'(t) + V_y \cdot y_0'(t)]$

Hence as in the proof of theorem27.2, we determine a solution of 27.6 such that for $(\xi,\eta)\varepsilon C$, $\bar{\varphi}(\xi_0(t), \eta_0(t)) \equiv z_0(t)$, $\bar{\varphi}_\xi \equiv \bar{p}_0(t)$ and $\bar{\varphi}_\eta \equiv \bar{q}_0(t)$; using theorem 28.2. The strip equation still holds under transformation. Then $\varphi(x,y) = \bar{\varphi}[U(x,y), V(x,y)]$ is the solution of (23.1) and $\varphi_x[x(t), y(t)] = \bar{\varphi}_\xi(\xi_0(t), \eta_0(t)) \cdot U_x(x_0(t), y_0(t)) + \bar{\varphi}_\eta \cdot U_y$ $= \bar{p}(t) \cdot U_x + \bar{q}(t) \cdot U_y \equiv p_0(t)$ and likewise $\varphi_y[x(t), y(t)] = q_0(t)$.

In the same way, we get from corollary to theorem (28.2), he solution of problem C_1 and I:

THEOREM 28.4: Given (a_0) and (a_1) as in theorem 28.3 and

(b_1) Let $x = \bar{x}_0(y)$ be a curve C where $\bar{x}_0(y)$ is of class A^1 in T.

(d_1) Let $\bar{z}_0(y)$ be of class A^1 in T and $\bar{p}_0(y)$ be of class A^0 in T.

Then there exists a function $\varphi(x,y)$ such that:

(α) $\varphi(x,y)$ is of class A^2 in $R \leq R_0$

(β) $z = \varphi(x,y)$ is a solution of (23.1) in R_0.

(δ_1) For $t \varepsilon T_1 \leq T$,

$$\varphi(\bar{x}_0(y),y) \equiv \bar{z}_0(y) \text{ and } \varphi_x(\bar{x}_0(y),y) \equiv \bar{p}_0(y).$$

THEOREM 28.5: Let (a_0) and (a_1) be given as in theorem 28.3.

(d_1) Let $\bar{z}_0(y)$ be of class A^1 in T and $\bar{p}_0(y)$ be of class A^0 in T.

Then there exists a function $\varphi(x,y)$ such that:

(α) $\varphi(x,y)$ is of class A^2 in $R \leq R_0$

(β) $z = \varphi(x,y)$ is a solution of (23.1) in R_0

(δ_0) For $t \varepsilon T_1 \leq T$,

$$\varphi(x_0,y) \equiv \bar{z}_0(y) \text{ and } \varphi_x(x_0,y) \equiv \bar{p}_0(y).$$

In Courant-Hilbert[1], v. II, pp. 326-332, theorem 28.3 is given with slightly weaker boundary conditions but with a different method of solution. It stems essentially from work of H. Lewy and will be discussed in Part D with the general hyperbolic equation.

COROLLARY: When (23.1) becomes (23.4) $\varphi(x,y) \equiv \bar{\varphi}(x + \frac{A}{\lambda_1} y, \ y + \frac{C}{\lambda_1} x)$

where $\bar{\varphi}(\xi,\eta)$ is the solution of (27.6) corresponding to the initial strip:

$$(28.6)\quad \begin{cases} \xi_0(t) = x_0(t) + \frac{A}{\lambda_1} y_0(t) \\ \eta_0(t) = y_0(t) + \frac{C}{\lambda_1} x_0(t) \\ \omega_0(t) = z_0(t) \\ \bar{p}_0(t) = p_0(t) \cdot \left(\frac{\lambda_1}{2\sqrt{B^2-AC}}\right) - q_0(t) \cdot \left(\frac{C}{2\sqrt{B^2-AC}}\right) \\ \bar{q}_0(t) = p_0(t) \cdot \left(\frac{-A}{2\sqrt{B^2-AC}}\right) + q_0(t) \cdot \left(\frac{\lambda_1}{2\sqrt{B^2-AC}}\right) \end{cases}$$

29. For the special equation: $\frac{\partial}{\partial x}\left\{p(x)\frac{\partial z}{\partial x}\right\} - \frac{\partial^2 u}{\partial y^2} = z^2$

with boundary conditions:

$$u(0,y) \equiv u(\pi,y) \equiv 0; \quad u(x,0) \equiv \bar{z}_0(x); \quad u_y(x,0) \equiv \bar{q}_0(x)$$

where $p(x) \geq 0$, Siddiqi[1] used a method due to Lichtenstein[1] to show that there is a unique solution of class A^2 in the rectangle

$$0 \leq x \leq \pi; \quad 0 \leq y \leq Y.$$

He carries through the proof by expanding $\bar{z}_0(x)$ and $\bar{q}_0(x)$ in series of characteristic functions of the system:

$$\begin{cases} \frac{d}{dx}\left(p(x)\frac{dy}{dx}\right) + \lambda y = 0 \\ y(0) = y(\pi) = 0 \end{cases}$$

gets a system of non-linear integral equations, which he solves by successive approximation.

Minakshi - Sundaram[1] generalized these methods to apply to:

$$(29.1)\quad \frac{\partial}{\partial x}\left[\, p(x)\frac{\partial z}{\partial x}\right] - \frac{\partial^2 z}{\partial y^2} = f(x,y,z,\frac{\partial z}{\partial x},\frac{\partial z}{\partial y})$$

but the paper was unavailable.

D. C. Lewis[1] considered the equation which is a special case of (29.1):

$$\frac{\partial^2 z}{\partial y^2} - \frac{\partial^2 z}{\partial x^2} + f(x,y,z,\frac{\partial z}{\partial x},\frac{\partial z}{\partial y}) = 0$$

where f is defined for all p and q, for $|z| < h$, $0 \leq y \leq Y$ and $0 \leq x \leq \pi$, and where f satisfies a Lipschitz condition in z, p, q, for $|z| \leq h$.

Boundary conditions are similar to Siddiqui's above:

$$z(0,y) = z(\pi,y) = 0;\ z(x,0) = f(x);\ z_t(x,0) = g(x)$$

Solution is by reduction to a trigonometric series where coefficients satisfy an infinite number of equations.

C. The Linear Hyperbolic Equation

30. Consider the equation:

$$(30.0)\quad a(x,y)\frac{\partial^2 z}{\partial x^2} + 2b(x,y)\frac{\partial^2 z}{\partial x\partial y} + c(x,y)\frac{\partial^2 z}{\partial z^2} + d(x,y)\frac{\partial z}{\partial x}$$
$$+ e(x,y)\frac{\partial z}{\partial y} + g(x,y)z = h(x,y) \qquad (ac<b^2)$$

The function $f(x,y,z,p,q) \equiv d(x,y)p + e(x,y)q + g(x,y)z - h(x,y)$ satisfies a Lipschitz condition whenever $(x,y)\varepsilon R$, a region where d, e, g, h are continuous, so that the theorems of Section C apply to (30.0) Under the transformation $\begin{cases}\xi = U(x,y)\\ \eta = V(x,y)\end{cases}$ of theorem 25.3 equation (30.0) reduces to:

$$(30.1) \quad \frac{\partial^2\omega}{\partial\xi\partial\eta} + \bar{d}(\xi,\eta)\cdot\frac{\partial\omega}{\partial\xi} + \bar{e}(\xi,\eta)\frac{\partial\omega}{\partial\eta} + \bar{g}(\xi,\eta)\omega = \bar{h}(\xi,\eta)$$

For such equations special methods have been developed. We therefore consider such equations first, giving the classical Riemann method.

Given

$$(30.2) \quad \frac{\partial^2 z}{\partial x\partial y} + d(x,y)\frac{\partial z}{\partial x} + e(x,y)\frac{\partial z}{\partial y} + g(x,y)z = h(x,y)$$

define the operators:

$$L(u) \equiv u_{xx} + du_x + eu_y + gu$$

$$M(v) \equiv v_{xx} - (dv)_x - (ev)_y + gv$$

The second is called the *adjoint* of the first. Now by theorem 27.1 there exists a solution of the Goursat problem:

$$M(v) = 0$$

$$v(\xi,y) = \tau(y) \equiv e^{\int_\eta^y d(\xi,\bar{y})d\bar{y}}$$

$$v(x,\eta) = \sigma(x) \equiv e^{\int_\xi^x e(\bar{x},\eta)d\bar{x}}$$

$(\tau(\eta) = \sigma(\xi = 1)$. This solution, corresponding to a given (ξ,η) is

$$v(x,y) = G(x,y;\xi,\eta)$$

and we call it the *Green's function* of (30.2).

Note that on the characteristic line $y = \eta$, $G(x,\eta;\xi,\eta) = \sigma(\xi)$, so $\frac{\partial G}{\partial x} - eG = 0$ and on the line $x = \xi$, $G(\xi,y;\xi,\eta) = \tau(\eta)$ so $\frac{\partial G}{\partial y} - dG = 0$, (using values of σ and τ). From this, we are able to get explicit solutions of problems G, C and I for (30.2).

THEOREM 30.1: (a) Let $\sigma(x)$, $\tau(y)$ be of class A^1 in S: $\alpha<x<\beta$, T: $\gamma<y<\delta$

(b) Let (x_0, y_0) be any point in the interior of $R = S \times T$ and suppose $\sigma(x_0) = \tau(y_0)$

(c_1) Let $d(x,y)$, $e(x,y)$ $g(x,y)$, $h(x,y)$ be of class A^1 in R

Then the function:

$$\varphi(x,y) = \sigma(x_0) \cdot G(x_0,y_0;x,y) + \int_{x_c}^{x}\int_{y_0}^{y} g(t,s) \cdot G(t,s;x,y)\,dt\,ds$$

$$+ \int_{x_0}^{x} [\,\sigma'(t) + e(t,y_0)\,\sigma(t)]\,G(t,y_0;x,y)\,dt$$

$$+ \int_{y_0}^{y} [\tau'(s) + d(x_0,s) \cdot \tau(s)]\;G(x_0,s;x,y)\,ds$$

(where G is the Green's function of (30.2), as defined on the previous page) is the unique function with the properties:

(α) $\varphi(x,y)$ is of class A^2 in R

(β) $z = \varphi(x,y)$ is a solution of (30.1) in R

(δ) $\varphi(x_0,y) = \tau(y)$ in T and $\varphi(x,y_0) = \sigma(x)$ in S.

Proof: Existence of solution $\bar{\varphi}(x,y)$ obtained in theorem 28.2. Hence, $L(\bar{\varphi}) = h$; also we know that $M(G) = 0$. By Green's theorem, (Kamke[1], pp. 411-412), if we take a fixed (ξ,η) and form a rectangle R which has (ξ,η) and (x_0,y_0) as opposite vertices, and boundary $\Gamma = C_1 + C_2 + C_3 + C_4$,

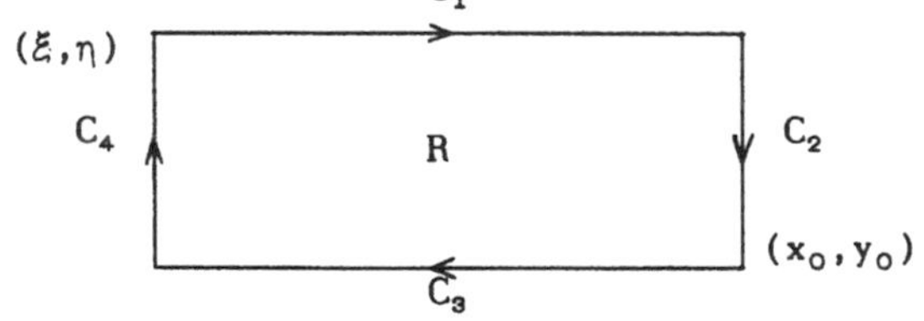

$$(30.3)\quad \iint_R G L(\bar{\varphi}) - \bar{\varphi}\,M(G)\quad dx\,dy = \int_\Gamma (P\,dy - Q\,dx)$$

where

$$P \equiv \tfrac{1}{2}[G \cdot \bar{\varphi}_y - \bar{\varphi} \cdot G_y] + d\bar{\varphi}G$$

and

$$Q \equiv \tfrac{1}{2}[G \cdot \bar{\varphi}_x - \bar{\varphi} \cdot G_x] + e\,\bar{\varphi}\,G$$

Since $M(G) = 0$ and $L(\bar{\varphi}) = h$, left side becomes:

$$\iint_R h(x,y) \cdot G(x,y;\xi,\eta)\,dx\,dy$$

Right side, if we evaluate it around all 4 sides ($dy = 0$ on C_1 and C_3, $dx = 0$ on C_2 and C_4) and use identities at the bottom of page 128, will become

$$\varphi(\xi,\eta) - \varphi(x_0,y_0)\,G\,(x_0,y_0;\xi,\eta) - \int_{x_0}^{\xi} G(x,y_0;\xi\,\eta)\,(\bar{\varphi}_x(x,y_0) + e\bar{\varphi})dx$$

$$- \int_{y_0}^{\eta} G(x_0,y;\xi,\eta)\,(\bar{\varphi}_y + d\bar{\varphi})dy$$

But on $y = y_0$, $\bar{\varphi} = \sigma(x)$ and $\bar{\varphi}_x = \sigma'(x)\cdot$ on $x = x_0$, $\bar{\varphi} = \tau(y)$ and $\bar{\varphi}_y = \tau'(y)$. Hence, substituting in (30.3), we get (30.2) except for notation (replace ξ, η by x, y, and change variables of integration).

THEOREM 30.2

(d) Let $\begin{aligned} x &= x_0(\mu) \\ y &= y_0(\mu) \end{aligned}$ be a curve C, where $x_0(\mu)$ and $y_0(\mu)$ are functions of class A^1 in M: $\mu_1<\mu<\mu_2$ and $x_0^{(1)}(\mu) \neq 0$, $y_0^{(1)}(\mu) \neq 0$ in M. Define α, β, γ, δ as in (b), page 119.

(a) Let $\sigma(x)$ be a function of class A^1 in S: $\alpha<x<\beta$
$\tau(y)$ be a function of class A^1 in T: $\gamma<y<\delta$.

(C_1) Let $d(x,y)$, $e(x,y)$, $g(x,y)$, $h(x,y)$ be of class A in $R=S\times T$.

Then, setting $s_1(x) = \mu$, the inverse of $x = x_0(\mu)$, $\mu = s_2(y)$ the inverse of $\bar{y} = y_0[s_1(x)]$ and $\bar{x} = x_0[s_2(y)]$, the function:

$$\varphi(x,y) = \tfrac{1}{2}[\sigma(x) + \tau(\bar{y})]\;G(x,\bar{y};x,y) + \tfrac{1}{2}[\sigma(\bar{x}) + \tau(y)]$$

$$(30.4)\quad G(\bar{x},y;x,y) + \int_{s_1(x)}^{s_2(y)} (P \cdot y_0'(\mu) - Q \cdot x_0'(\mu))\,d\mu +$$

$$\iint_{R_{xy}} h(t,s)\;G(t,s;x,y)\,dt\,ds$$

where G is the Green's function defined on page 128 and

$$P = \tfrac{1}{2}G(x_0(\mu), y_0(\mu); x, y) \cdot \tau'(y_0(\mu)) + [\sigma(x_0(\mu)) + \tau(y_0(\mu))]$$

$$[dG - \tfrac{1}{2}G_y]_{x = x_0(\mu),\ y = y_0(\mu)}$$

$$Q = \tfrac{1}{2}G(x_0(\mu), y_0(\mu); x, y) \cdot \sigma'(x_0(\mu)) + [\sigma(x_0(\mu)) + \tau(y_0(\mu))]$$

$$[dG - \tfrac{1}{2}G_x]_{x = x_0(\mu),\ y = y_0(\mu)}$$

is the unique function such that

(α) $\varphi(x,y)$ is of class A^2 in R

(β) $z = \varphi(x,y)$ satisfies (30.2) in R

(γ) For $\mu \varepsilon M$, $\varphi(x_0(\mu), y_0(\mu)) = \sigma(x_0(\mu)) + \tau(y_0(\mu))$.

Proof: Use Green's theorem again, with Γ the complete boundary of R, and as in the previous theorem

$$\iint_{R_{xy}} h(t,s)\,dt\,ds = \iint_{R_{xy}} (GL(\varphi) - \varphi M(G)) = \int_{\Gamma} [P\,dy - Q\,dx]$$

THEOREM 30.3: Given (d) and (c_1) as above by (b) replaced by:

($\overline{G}$) Let $z_0(\mu)$, $p_0(\mu)$, $q_0(\mu)$ be defined in M, where $z_0'(\mu)$, $p_0(\mu)$, $q_0(\mu)$ are continuous and $z_0'(\mu) = p_0(\mu) \cdot x_0'(\mu) + q_0(\mu) \cdot y_0'(\mu)$

Then the unique function $\varphi(x,y)$ satisfying (α), (β) and:

($\overline{\gamma}$) For $\mu \varepsilon M$, $\varphi(x_0(\mu), y_0(\mu)) = z_0(\mu)$; $\varphi_x(x_0(\mu), y_0(\mu)) = p_0(\mu)$; $\varphi_y(x_0(\mu), y_0(\mu)) = q_0(\mu)$

is given by (30.3) where

$$\sigma(x) = \int_{x_0}^{x} p(s_1(t))\,dt + \frac{z_0}{2};\ \tau(y) = \int_{y_0}^{y} q(s_2(t))\,dt + \frac{z_0}{2}$$

$$(30.5) \quad \text{or } \varphi(x,y) = \tfrac{1}{2}z_0(s_1(x)) \cdot G(x,\overline{y};x,y) + \tfrac{1}{2}z_0(s_2(y)) \cdot G(\overline{x},y;x,y) + \int_{\Gamma_1} \ldots + \iint_R \ldots$$

31. In one important case, $G(x,y;\xi,\eta)$ can be explicitly given.

$$(31.1) \quad \frac{\partial^2 z}{\partial x \partial y} + D\frac{\partial z}{\partial x} + E\frac{\partial z}{\partial y} + Gz = h(x,y) \quad (D,E,G \text{ are constants})$$

the Green's function is the solution of

$$v_{xy} - Dv_x - Ev_y + Gv = 0$$

such that on lines $x = \xi$ and $y = \eta$ this assumes values

$$\tau(y) = e^{D(y-\eta)} \text{ and } \sigma(x) = e^{E(x-\xi)}$$

respectively. If we set

$$w = ve^{-(Dy + Ex)}$$

then direct substitution in the equation and boundary conditions shows that we need a solution for

$$(31.2) \quad \begin{cases} w_{xy} - [G - ED]\, w = 0 \\ w(\xi,y) = w(x,\eta) = e^{-(D\eta + E\xi)} \quad \text{(a constant)} \end{cases}$$

The solution of:

$$(31.3) \quad \begin{cases} \dfrac{\partial^2 z}{\partial x \partial y} + \bar{G}\, z = 0 \\ \tau(y) = z(\xi,y) = 1 \text{ and } \sigma(x) = z(x,\eta) = 1 \end{cases}$$

reduces to an ordinary differential equation. It is:

$$(31.4) \quad G_0(x,y;\xi,\eta) = \sum_{n=0}^{\infty} \frac{(-1)^n}{(n!)^2} [\bar{G}(x-\xi)(y-\eta)]^n = J_0\left(\sqrt{-2\,\bar{G}(x-\xi)(y-\eta)}\right)$$

Thus solution of (31.2) is $w = e^{-(D\eta + E\xi)} \cdot G_0(x,y;\xi,\eta)$ with $\bar{G} = G\text{-}DE$.

Hence Green's function of original problem would be $v = we^{Dy + Ex}$ or

$$(31.5) \quad G(x,y;\xi,\eta) = e^{-D(y + \eta) + E(x + \xi)]} \cdot G_0(x,y;\xi,\eta)$$

where $\bar{G} = G\text{-}DE$, and G_0 is given in 31.4.

32. Returning now to the general equation (30.0), we have:

LEMMA: Under the transformation $\begin{cases} \xi = U(x,y) \\ \eta = V(x,y) \end{cases}$ of theorem (25.3), equation (30.0) becomes (30.1) where

$$(32.1) \quad \begin{cases} \bar{d}\,(\xi,\eta) = \dfrac{-\bar{L}\,(U)}{2J\,(b^2-ac)}, & \text{on substituting } x = X(\xi,\eta),\ y = Y(\xi,\eta) \text{ on the right side} \\ \bar{e}\,(\xi,\eta) = \dfrac{-\bar{L}\,(V)}{2J\,(b^2-ac)} & \text{''} \\ \bar{g}\,(\xi,\eta) = \dfrac{-g}{2J\,(b^2-ac)} & \text{''} \\ \bar{h}\,(\xi,\eta) = \dfrac{-h}{2J\,(b^2-ac)} & \text{''} \end{cases}$$

where $\bar{L}(z) \equiv a.\ z_{xx} + 2bz_{xy} + cz_{yy} + dz_x + ez_y$.

Proof: If we define X and Y as above, and let $p = \bar{p}\,U_x + \bar{q}\,V_x$, $q = \bar{q}\,V_y + \bar{p}\,U_y$, then the formulas on page 98 become, since $f(x,y,z,p,q) = dp+eq+gz+h$, and $\alpha = \gamma = 0$,

$$2\beta\bar{\varphi}_{\xi\eta} + \delta\bar{\varphi}_{\xi} + \varepsilon\bar{\varphi}_{\eta} + (U_x\bar{\varphi}_{\xi} + V_x\bar{\varphi}_{\eta})d + (U_y\varphi_{\xi} + V_y\bar{\varphi}_{\eta})e + g\bar{\varphi} + h$$
$$= 2\beta\bar{\varphi}_{\xi\eta} + [\delta + dU_x + eU_y]\,\bar{\varphi}_{\xi} + [\varepsilon + dV_x + dV_y]\,\bar{\varphi}_{\eta} + g\bar{\varphi} + h$$
$$= -2\sqrt{b^2-ac}\;J\cdot\bar{\varphi}_{\xi\eta} + [aU_{xx} + 2bU_{xy} + cU_{yy} + dU_x + eU_y]\,\bar{\varphi}_{\xi}$$
$$+[aV_{xx} + 2bV_{xy} + cV_{yy} + dV_x + eV_y]\,\bar{\varphi}_{\eta} + g\bar{\varphi} + h = 0$$

Dividing by coefficient of $\varphi_{\xi\eta}$, which is not 0, we get coefficients defined as above: so it becomes

$$\bar{\varphi}_{\xi\eta} + \bar{d}(\xi,\eta)\cdot\bar{\varphi}_{\xi} + \bar{e}(\xi,\eta)\cdot\bar{\varphi}_{\eta} + \bar{g}(\xi,\eta)\cdot\bar{\varphi} + \bar{h}(\xi,\eta) = 0$$

Thus $w = \bar{\varphi}(\xi,\eta)$ is a solution of (30.1) if $z = \varphi(x,y)$ is a solution of (30.0) and conversely.

THEOREM 32.1 Let $a(x,y)$, $b(x,y)$, $c(x,y)$ be functions of class A^2 and $d(x,y)$, $\ldots, h(x,y)$ be continuous in φ_0. Then the function $\varphi(x,y)$ of Theorem 27.2

(solution to problem G) exists in every closed bounded subregion $R_1 \leq R_0$ when (23.1) is of the linear form (30.0). It is obtained as follows:

(1) Determine $\begin{cases} \xi = U(x,y) \\ \eta = V(x,y) \end{cases}$ of Theorem 25.3. For given $\begin{cases} x(\mu), y(\mu), \\ \bar{x}(\nu), \bar{y}(\nu) \end{cases}$

let $h(\mu) \equiv U(x(\mu), y(\mu))$ and $k(\nu) \equiv V(\bar{x}(\nu), \bar{y}(\nu))$. Let $\mu = s_1(\eta)$ be the inverse of $\eta = h(\mu)$ and let $\nu = s_2(\xi)$ be the inverse of $\xi = k(\nu)$.

(2) Let $\xi_0 = U(x_0, y_0)$ and $\eta_0 = V(x_0, y_0)$; let $F(\eta) = \sigma(s_1(\eta))$ and $\bar{\sigma}(\xi) = \tau(s_2(\xi))$. Write equation (30.1) where coefficients are defined in (32.1).

(3) Find the Green's function $G(\xi,\eta;\bar{\xi},\bar{\eta})$ of equation (30.1). Using this get solution $w = \bar{\varphi}(\xi,\eta)$ of problem G for (30.1) - i.e. use formula (30.4) with $\bar{\sigma}(\xi)$ and $\bar{\tau}(\eta)$ for $\sigma(x)$ and $\tau(y)$.

(4) Let $\varphi(x,y) = \bar{\varphi}(U(x,y), V(x,y))$. Then $z = \varphi(x,y)$ is required solution.

THEOREM 32.2: Let $a(x,y)$, $b(x,y)$, $c(x,y)$ be of class A^2 and $d(x,y), \ldots, h(x,y)$ be continuous in R_0. Then the function $\varphi(x,y)$ of theorem 28.2 (problem C) exists for every closed bounded $R \leq R_0$. It is obtained as follows:

(1) Determine $\begin{cases} \xi = U(x,y) \\ \eta = V(x,y) \end{cases}$ of theorem 4.3 and let $\begin{cases} \xi_0(t) \equiv U[x_0(t), y_0(t)] \\ \eta_0(t) \equiv V[x_0(t), y_0(t)] \end{cases}$

and $\begin{matrix} \bar{p}_0(t) = p(t) \cdot X_\xi + q(t) \cdot Y_\xi \\ \bar{q}_0(t) = p(t) \cdot X\eta + q(t) \cdot Y\eta. \end{matrix}$ Let $\begin{cases} \xi_0 = \xi_0(t_0). \\ \eta_0 = \eta_0(t_0). \end{cases}$

(2) Write equation (30.1) where coefficients are given in (31.1) and find its Green's function. Use this to get solution $\omega = \bar{\varphi}(\xi,\eta)$ of problem C for (30.1) - i.e. use formula (30.5).

(3) Let $\varphi(x,y) = \bar{\varphi}(U(x,y), V(x,y))$. This is required solution.

33 When $a(x,y) = A$, $b(x,y) = B$, $c(x,y) = C$, U_{xx}, U_{xy}, U_{yy} are all $= 0$, and by direct substitution

$$\bar{d}(\xi,\eta) = \frac{-1}{2J(B^2 \ -AC)} \left[d + \frac{A}{\lambda_1} \right]_{x = X(\xi,\eta);\ y = Y(\xi,\eta)}$$

$$\bar{e}(\xi,\eta) = \frac{-1}{2J(B^2 \ -AC)} \left[e + \frac{C}{\lambda_1} \right]_{x = X(\xi,\eta);\ y = Y(\xi,\eta)}$$

In particular, if $d = D$, $e = E$, $g = G$, where these are constants, (30.1) is

$$\frac{\partial^2 \omega}{\partial\xi\partial\eta} + \frac{1}{2J\sqrt{b^2 - ac}} \left[\left(D + \frac{A}{\lambda_1} E\right) \frac{\partial\omega}{\partial\xi} + \left(E + \frac{C}{\lambda_1} D\right) \frac{\partial\omega}{\partial\eta} + g\omega \right] = h(\xi,\eta)$$

so that it is always possible, by using explicit formulations for Green's functions given on page 132, to get solutions of problems G, C, $\bar{C}$, I.

$$(33.1) \quad A\frac{\partial^2 z}{\partial x^2} + 2B\frac{\partial^2 z}{\partial x\partial y} + C\frac{\partial^2 z}{\partial y^2} + D\frac{\partial z}{\partial x} + E\frac{\partial z}{\partial y} + Gz = h(x,y)$$

34. Consider the equation:

$$(34.1) \quad \frac{\partial^2 z}{\partial x^2} - \frac{\partial^2 z}{\partial y^2} = 0$$

Characteristic curves are

$$\Gamma_1: \ (x-x_0) = -(y-y_0)$$

$$\Gamma_2: \ (y-y_0) = \ (x-x_0)$$

Thus applying transformation $\begin{cases} \xi = U(x,y) \equiv x+y \\ \eta = V(x,y) \equiv y-x \end{cases}$ we get $\dfrac{\partial^2 \omega}{\partial\xi\partial\eta} = 0.$

Now the solutions of problems G, C, $\bar{C}$, I for:

$$(34.2) \quad \frac{\partial^2 z}{\partial x\partial y} = 0$$

are easily obtained. Solution to problem G is:

$$\varphi(x,y) = \sigma(x) + \tau(y) - \sigma(x_0)$$

Solution to $\bar{C}$ is:

$$\varphi(x,y) = \sigma(x) + t(y)$$

Solution to C (Cauchy) is

$$\varphi(x,y) = \frac{z\,(s_1(x)) + z(s_2(y))}{2} + \tfrac{1}{2}\int_{s_1(x)}^{s_2(y)} \cdot [p_0'(\mu) \cdot x'(\mu)$$

$$- q_0'(\mu) \cdot y'(\mu)]\, d\mu$$

$$= z(\mu_0) + \int_{\mu_0}^{s_1(x)} p_0'(\mu) \cdot x_0'(\mu)\, d\mu + \int_{\mu_0}^{s_2(y)} q_0'(\mu)$$

$$\cdot\, y_0'(\mu)\, d\mu$$

In particular, when the curve is $\begin{array}{l} x = x_0 \\ y = \mu \end{array}$ - (problem I), we get

$$\varphi(x,y) = \int_{y_0}^{y} q_0'(\mu)\, d\mu + \bar{z}(y_0).$$

If curve is $\begin{cases} y = y_0 \\ x = \mu \end{cases}$ solution would be $\mu(x,y) = \bar{\bar{z}}(x_0) + \int_{x_0}^{x} p_0'(\mu)\, d\mu.$

Hence using the above transformation we can easily get solutions of (34.1).

Thus for I,

$$\varphi(x,y) = \bar{z}(y_0 - x_0) + \int_{x-y}^{x+y} q_0^{(1)}(\mu)\, d\mu.$$

However, there are other methods which can be used also to solve (34.1). As an example of the method of finite differences, as applied to these problems, take problem I_0 for:

$$(34.3)\quad \frac{\partial^2 u}{\partial x^2} - \frac{\partial^2 u}{\partial y^2} = 0$$

(Courant, Freidrichs, Lewy[1]). We wish to find a solution of (34.1), $z = \varphi(x,y)$ such that

$$\varphi(x,0) = \bar{z}_0(x)$$

$$\varphi_y(x,0) = \bar{q}_0(x) = \bar{z}_0'(x) + 2\, q_0(x).$$

where $\bar{z}_0(x)$ and $\bar{q}_0(x)$ are functions of class A^2 and A^1 respectively.

Lay off a square lattice: $x = nh,\ y = mh$ $(n,m = 0, \pm 1, \pm 2, \ldots,)$

and define finite differences:

$$\frac{1}{h}[u(x+h,y) - u(x,y)] = u_x \ ; \quad \frac{1}{h}[u(x,y+h) - u(x,y)] = u_y$$

$$\frac{1}{h}[u(x,y) - u(x-h,y)] = u_{\bar{x}}; \quad \frac{1}{h}[u(x,y) - u(x,y-h)] = u_{\bar{y}}$$

Set $u_{x\bar{x}} = u_{\bar{x}x} = (u_x)_{\bar{x}} = \frac{1}{h^2}[u(x+h,y) - 2u(x,y) + u(x-h,y)]$

Then replace (34.3) by the difference equation

$$(34.4) \quad u_{y\bar{y}} - u_{x\bar{x}} = 0$$

Given any net point $P_0 = (x_0,y_0)$, call value of u there u_0; likewise define:

$$u_1 = u(x_0,y_0+h)$$

$$u_2 = u(x_0+h,y_0)$$

$$u_3 = u(x_0,y_0-h)$$

$$u_4 = u(x_0-h,y_0)$$

Then the difference equation (34.4) becomes:

$$(34.5) \quad u_1 + u_3 - u_2 - u_4 = 0$$

Note that P_0 has not entered into (34.5); indeed the equation (34.5) only converts values of u at alternate points of the mesh; so we have two sets S and $\bar{S}$, of values, and we can confine ourselves to one subset, say S. Now we are given the values of u at mesh-points on the line $y = 0$, the lowest line, and since we know u_y, we can write (taking it as a difference)

$$u_y(nh,0) = \frac{1}{h}[u(nh,h) - u(nh,0)],$$

so we can get $u(nh,h)$. That is, initial conditions give values on lowest two lines of the mesh. But then by (34.5), the values of u at any point in the third row is given, in terms of its three neighboring points in the two bottom rows: $u_1 = u_2 - u_3 + u_4$. Proceeding in this way, we can get the value of u uniquely determined at every point of S. Furthermore, it is apparent that for a given point P, its value is determined by the values on the first two lines between two numbers α and β obtained by drawing lines $x \times y$ = constant and $x-y$ = constant through P. Now if we denote:

$$u_1^{\swarrow} = u_1 - u_4 \qquad u_1^{\searrow} = u_1 - u_2$$

$$u_2^{\swarrow} = u_2 - u_3 \qquad u_4^{\searrow} = u_4 - u_3$$

where slant of arrow corresponds to direction along characteristic, (34.5) becomes

$$u_1^{\swarrow} = u_2^{\swarrow}.$$

So differences along any one characteristic line, in direction of the other characteristic are all equal, and hence equal to one of the *given* differences on the two bottom rows. On the other hand, $u_p - u_\alpha$ = sum of all differences along line $P\alpha$, so we get

$$(34.6) \quad u_p = u_\alpha + \sum_\alpha^\beta u^{\swarrow}$$

But now, if we want a formula for solution of (34.3),, let $h \to 0$, where given values on bottom line $\to \bar{z}_0(x)$ while the difference quotients $\frac{u^{\swarrow}}{h\sqrt{2}} \to \bar{q}_0(x)$. The right side of (34.6) then $\to$

$$\bar{z}_0(x-y) + \frac{1}{\sqrt{2}} \int_{x-y}^{x+y} \bar{q}_0(\xi)\, d\xi,$$

and left side $\to \varphi(x,y)$. Hence we get

$$\varphi(x,y) = \bar{z}_0(x-y) + \frac{1}{\sqrt{2}} \int_{x-y}^{x+y} \bar{q}_0(\xi)\, d\xi$$

where initial conditions are as given.

35. In a similar manner, one could find solutions of:

$$(35.1) \quad \frac{\partial^2 u}{\partial x^2} - k^2 \frac{\partial^2 u}{\partial y^2} + \alpha \frac{\partial u}{\partial x} + \beta \frac{\partial u}{\partial y} + \gamma u = 0$$

In section A, it was shown that a homogeneous linear hyperbolic equation with constant coefficients could always be reduced to the above form, and indeed to the form: $z_{xx} - z_{yy} + az = 0$.

Chandrasekar[1] considered equation:

$$\frac{\partial^2 z}{\partial x^2} - \frac{\partial^2 z}{\partial y^2} + z = 0$$

under conditions:

$$z(x,0) = \bar{z}_0(x); \quad z_y(x,0) = \bar{q}_0(x) \qquad (0 \le x \le 1)$$

and

$$z(0,y) = \varphi(y) \qquad (y \ge 0)$$

$$z(1,y) = \psi(y) \qquad (y \ge 0)$$

He obtained solution by constructing a function $G_0(x,y;\xi,\eta)$ which is 1 on characteristic $x+y = \xi+\eta$ and on line $x = \xi$, $y \le \eta$. Then by a Green's identity, and straight Riemann method, he obtained a solution involving only quadratures.

Bourgin and Duffin[1] considered the analogue of Dirichlet and Neumann problems for the vibrating string equation (34.1). They established results of the following nature:

Let R be a rectangle: $0 \le x \le S$, $0 \le y \le Y$ and let $\alpha = \frac{Y}{S}$.

Let $\varphi(x,y)$ be a solution of (34.1) in the interior of R which vanishes on the closed boundary; let $\varphi(x,y)$ be of class A^1 in R and φ_{xx} and φ_{yy} be Lebesgue summable in R. Then in order that the solution be unique, it is necessary and sufficient that α be irrational.

36. Various other problems have been studied for equation (35.1). Hamburger[1] has considered the homogeneous equation $[h(x,y) \equiv 0]$ where the coefficients d, e, g are periodic functions of y in a strip $x_0 \leq x \leq x_1$ $d(x,y+2\pi) = d(x,y)$, $e(x,y+2\pi) = e(x,y)$, $g(x,y+2\pi) = g(y)$ and tried to find an integral $\varphi(x,y)$ with properties (α), (β) and:

$$(k) \quad \varphi(x,y+2\pi) = \varphi(x,y) \quad \text{for } x_0 \leq x \leq x_1.$$

He has obtained, under rather strong assumptions concerning the behavior of the coefficients, conditions for a solution, and seems to have carried out an approximation method for obtaining the solution.

Badescu[1] worked out a simplification of a method due to Picone that could be used to solve various problems such as G for equation (35.1) and for the slightly more general one obtained by replacing $g(x,y)z - h(x,y)$ by $m(x,y,z)$.

Einaudi[1] considering a simple case in which the coefficients are singular:

$$(36.1) \quad d(x,y) = \frac{\alpha(x,y)}{x-y}, \quad e(x,y) = \frac{\beta(x,y)}{x-y}, \quad g(x,y) = \frac{\gamma(x,y)}{x-y},$$

$$h(x,y) = (x-y)^h\, p(x,y) \qquad (h \geq 0) \quad , \quad k = 1 \quad ,$$

established, by a method of successive approximation, the following:

If α, β, γ are continuous in the triangle Δ indicated,

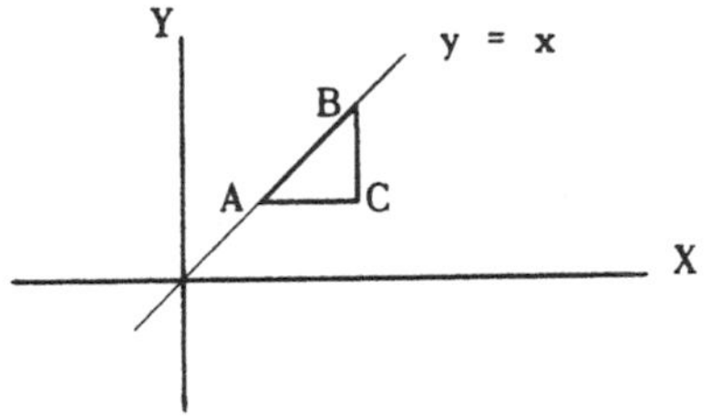

and if $\alpha \leq m$, $\beta \leq n$, $\gamma \leq p$, and we set

$$\rho = \frac{m}{k+1} + \frac{n}{k+1} + \frac{p}{(k+1)(k+2)}$$

then if $\rho < 1$, there exists a function $\varphi(x,y)$ of class A^1 in Δ such that $z = \varphi(x,y)$ satisfies (35.1), with coefficients given in (36.1), and such that $(x-y)\ k+2\ \varphi(x,y) \to 0$

$$(x-y)\ k+1\ \varphi_x(x,y) \to 0$$

$$(x-y)\ k+1\ \varphi_y(x,y) \to 0$$

as $(x,y) \to$ a point on AB. It is not clear how the limit is taken; probably normally.

Bergmann[1] solved the problem of Cauchy by relating the unknown functions to a special kind of complex operator, and getting approximating polynomials. Ingersoll[1] solved a problem analogous to the Goursat problem for equation (35.1) except that the initial data was the value of u_{x_k} on the line $x = 0$ and u_{y_n} on the line $y = 0$, rather than the values of u itself.

D. General Hyperbolic Equation

$$(22.2)\quad F\left(x,y,z,\frac{\partial z}{\partial x},\frac{\partial z}{\partial y},\frac{\partial^2 z}{\partial x^2},\frac{\partial^2 z}{\partial x \partial y},\frac{\partial^2 z}{\partial y^2}\right) = 0 \qquad (F_r F_t < F_s^{\,2})$$

37. In 1926 Lewy[1] obtained a general solution to problem C, by using systems of first order partial differential equations, which were in turn solved by finite differences. Instead of taking the initial curve in terms of a single parameter μ, however, he assumes a plane, with coordinates (α,β)

considers a line segment in that plane:

$$L: \quad \alpha + \beta = 0, \ \alpha \leq \alpha_0, \ \beta \leq \beta_0, \text{ where } \alpha_0 + \beta_0 > 0$$

and assumes that we have $x(\alpha,\beta)$, $y(\alpha,\beta)$, $z(\alpha,\beta)$, $p(\alpha,\beta)$, $q(\alpha,\beta)$, $r(\alpha,\beta)$, $s(\alpha,\beta)$, $t(\alpha,\beta)$ given along this line in terms of arc length; that is, if λ is a parameter, equations of L are:

$$L: \quad \alpha = \frac{-\lambda}{\sqrt{2}}, \ \beta = \frac{+\lambda}{\sqrt{2}} \text{ for } \lambda \varepsilon M: \lambda_1 \leq \lambda \leq \lambda_2 \text{ where } \lambda_1 < 0 < \lambda_2.$$

and we are given $x_0(\lambda) \equiv x\left(\frac{-\lambda}{\sqrt{2}}, \frac{\lambda}{\sqrt{2}}\right)$, $y_0(\lambda), \ldots, t_0(\lambda)$ such that these eight elements form a strip of second order, (i.e., satisfy equations (24.4) and such that

$$\Delta \equiv \begin{vmatrix} F_r & F_s & F_t \\ x_0^1(\lambda) & y_0^1(\lambda) & 0 \\ 0 & x_0^1(\lambda) & y_0^1(\lambda) \end{vmatrix} \neq 0$$

where in F_r, F_s, F_t we have substituted $x = x_0(\lambda), \ldots, t = t_0(\lambda)$, which means that the strip is not a characteristic strip((24.5) does not hold). By analogy to (25.1), in the quasi-linear case, we can call:

$$(37.1) \quad Ry^{(1)2} - Sx^{(1)}y^{(1)} + Tx^{(1)2} = 0$$

the *characteristic equation* of (22.2), where $R \equiv F_r$, $S \equiv F_s$, $T \equiv F_t$. Assume that $F_r \neq 0$ and $F_t \neq 0$. (Lewy says that by a rotation of the x-y plane, this can always be assumed to hold; it seems to me that such a rotation can be performed in the neighborhood of a point but not necessarily for the entire plane region) Because $4RT - S^2 = \Delta < 0$, equation (37.1) has two real linear factors:

$$(\rho_1 x' - \sigma_1 y\alpha)(\rho_2 x' - \sigma_2 y') = 0$$

where $\rho_1\rho_2 = T$, $\sigma_1\sigma_2 = R$, $\rho_1\sigma_2 + \rho_2\sigma_1 = S$ and hence $0 < -\Delta = (\rho_1\sigma_2 - \rho_2\sigma_1)^2$. Now if $\sigma_1 = 1$, $\sigma_2 = R \neq 0$ and we get

$$\rho_1 = \tfrac{1}{2}[S - \sqrt{-\Delta}], \ \rho_2 = \frac{1}{2R}[S + \sqrt{-\Delta}] \text{ so } \rho_i \neq 0 \ (i = 1, 2).$$

Thus we can solve:

$$(37.2) \quad \rho_1 x^1 - \sigma_1 y^1 = 0$$

$$\rho_2 x^1 - \sigma_2 y^1 = 0$$

Let the characteristics be the curves α = constant and β = constant. Then we have, from the above equations:

$$(37.3) \quad \rho_1 \frac{\partial x}{\partial \alpha} - \sigma_1 \frac{\partial y}{\partial \alpha} = 0$$

$$\rho_2 \frac{\partial x}{\partial \beta} - \sigma_2 \frac{\partial y}{\partial \beta} = 0$$

[That is, although it is not explicitly stated, we get two distinct families of solutions of (22.1): $\alpha(x,y)$ = constant and $\beta = (x,y)$ = constant. Then make a transformation of variables: $\alpha = \alpha(x,y)$ and $\beta = \beta(x,y)$. This has inverse $x = x(\alpha,\beta)$, $y = y(\alpha,\beta)$ under which equation (22.2) gets transferred and also original curve C goes into straight line L. Now for any curve in (α,β) plane: $\alpha = f(\mu)$, $y = g(\mu)$, first equation of (37.2) becomes $\rho_1 [\frac{\partial x}{\partial \alpha}\frac{d\alpha}{d\mu} + \frac{\partial x}{\partial \beta}\frac{d\beta}{d\mu}] - \sigma_1 [\frac{\partial y}{\partial \alpha}\frac{d\alpha}{d\mu} + \frac{\partial y}{\partial \beta}\frac{d\beta}{d\mu}] = 0$. However, on curve α = constant, taking $\alpha = \mu$, $\frac{d\alpha}{d\mu} = 1$ and $\frac{d\beta}{d\mu} = 0$, so we get first of equations (37.3). Likewise, on other characteristic β = constant, and equation (37.2) becomes second equation of (37.3)] The other equations of the characteristic strips (24.5) must also hold. These become (see Lewy[1] and Friedrichs and Lewy[1]), using notation of (24.5):

$$(37.4) \quad \begin{cases} \rho_1 R \frac{\partial r}{\partial \alpha} + \sigma_1 T \frac{\partial s}{\partial \alpha} + \sigma_1 X \frac{\partial y}{\partial \alpha} = 0 \\ \rho_1 R \frac{\partial s}{\partial \alpha} + \sigma_1 T \frac{\partial t}{\partial \alpha} + \sigma_1 Y \frac{\partial y}{\partial \alpha} = 0 \\ \frac{\partial z}{\partial \alpha} - p \frac{\partial x}{\partial \alpha} - q \frac{\partial y}{\partial \alpha} = 0 \end{cases}$$

$$(37.4)\quad \begin{cases} \dfrac{\partial p}{\partial\alpha} - r\dfrac{\partial x}{\partial\alpha} - s\dfrac{\partial y}{\partial\alpha} = 0 \\[2ex] \dfrac{\partial q}{\partial\alpha} - s\dfrac{\partial x}{\partial\alpha} - t\dfrac{\partial y}{\partial\alpha} = 0 \\[2ex] \rho_2\, R\dfrac{\partial r}{\partial\beta} + \sigma_2\, T\dfrac{\partial s}{\partial\beta} + \sigma_2\, X\dfrac{\partial y}{\partial\beta} = 0 \end{cases}$$

(There are other equations, on lines β = constant, we have not used.) These eight equations can be solved to get x, y, z, p, q, r, s, t as functions of α and β, by theorem 19.5, provided determinant of coefficients is not 0. In notation of that theorem, taking $m=8$, $\varphi_1 = x$, $\varphi_2 = y$, $\varphi_3 = r$, $\varphi_4 = s$, $\varphi_5 = t$, $\varphi_6 = z$, $\varphi_7 = p$, $\varphi_8 = q$.

$$\bar{\Delta} = \begin{vmatrix} \rho_1 & -\sigma_1 & 0 & 0 & 0 & 0 & 0 & 0 \\ 0 & \sigma_1 x & \rho_1 R & \rho_1 T & 0 & 0 & 0 & 0 \\ 0 & \sigma_1 Y & 0 & \rho_1 R & \sigma_1 T & 0 & 0 & 0 \\ -p & -q & 0 & 0 & 0 & 1 & 0 & 0 \\ -r & -s & 0 & 0 & 0 & 0 & 1 & 0 \\ -s & -t & 0 & 0 & 0 & 0 & 0 & 1 \\ 00 & \sigma_2 X & \rho_2 R & \sigma_2 T & 0 & 0 & 0 & 0 \\ \rho_2 & -\sigma_2 & 0 & 0 & 0 & 0 & 0 & 0 \end{vmatrix} = - \begin{vmatrix} \rho_1 & -\sigma_1 & 0 & P & 0 & 0 & 0 & 0 \\ \rho_2 & -\sigma_2 & 0 & 0 & 0 & 0 & 0 & 0 \\ 0 & \sigma_1 X & \rho_1 R & \sigma_1 T & 0 & 0 & 0 & 0 \\ 0 & \sigma_1 Y & 0 & \rho_1 R & \sigma_1 T & 0 & 0 & 0 \\ 0 & \sigma_2 X & \rho_2 R & \sigma_2 T & 0 & 0 & 0 & 0 \\ -p & -q & 0 & 0 & 0 & 1 & 0 & 0 \\ -r & -s & 0 & 0 & 0 & 0 & 1 & 0 \\ -s & -t & 0 & 0 & 0 & 0 & 0 & 1 \end{vmatrix}$$

Expanding: $\bar{\Delta} = -[\rho_1\sigma_2 - \rho_2\sigma_1]^2\, \sigma_1\, RT^2 = \Delta\, F_r\, F_t \neq 0 \quad (\sigma_1 = 1)$.

Now given initial values $x_0(\lambda),\ldots,z_0(\lambda)$ on line $\alpha + \beta = 0$, by theorem 19.5, there is a unique solution of (37.3) and (37.4).

$$x = x(\alpha,\beta),\ y = y(\alpha,\beta),\ z = z(\alpha,\beta),\ p = p(\alpha,\beta),\ q = q(\alpha,\beta),\ r = r(\alpha,\beta),$$
$$s = s(\alpha,\beta),\ t = t(\alpha,\beta)$$

in the neighborhood of line L which takes on prescribed boundary values. Furthermore, from first two equations (37.3)

$$\frac{\dfrac{\partial y}{\partial\alpha}}{\dfrac{\partial x}{\partial\alpha}} - \frac{\dfrac{\partial y}{\partial\beta}}{\dfrac{\partial x}{\partial\beta}} = \frac{\rho_1}{\sigma_1} - \frac{\rho_2}{\sigma_2} \neq 0,$$

so we can solve and get α and β as functions of x and y, and therefore get

(37.5) $z = z(\alpha(x,y),\beta(x,y)) \equiv \varphi(x,y),\ p = \bar{p}(x,y),\ldots,t = \bar{t}(x,y).$

Since (37.5) can be shown to satisfy

$$F(x,y,z,p,q,r,s,t) = 0$$

as well as the other strip equations

$$\frac{\partial z}{\partial \beta} - p\frac{\partial x}{\partial \beta} - q\frac{\partial y}{\partial \beta} = 0,\ \frac{\partial p}{\partial \beta} - r\frac{\partial x}{\partial \beta} - s\frac{\partial y}{\partial \beta} = 0,\ \frac{\partial q}{\partial \beta} - s\frac{\partial x}{\partial \beta} - t\frac{\partial y}{\partial \beta} = 0,$$

this means that we can take $z = \varphi(x,y)$ and $p = \varphi_x(x,y)$ and $q = \varphi_y(x,y)$ and $r = \varphi_{xx}(x,y)$, $s= \varphi_{xy}(x,y)$ and $t = \varphi_{yy}(x,y)$. Hence $z = \varphi(x,y)$ is a solution of equation (22.2) and has prescribed boundary conditions. In other words, Lewy established the following: (compare theorem 28.3, page 123)

THEOREM 37.1:

(a) Let $F(x,y,z,p,q,r,s,t)$ be a function of class A^3 and let $\Delta = 4F_r\, F_t - F_s^{\,2} < 0$ everywhere in a region S. Let $F_r \neq 0$ and $F_t \neq 0$ in S. Let the solutions of the characteristic equation (37.1) be $\alpha(x,y)$ = constant and $\beta(x,y)$ = constant and set $\alpha = \alpha(x,y)$, $\beta = \beta(x,y)$.

(b) Let C: $\begin{cases} x = x_0(\mu) \\ y = y_0(\mu) \end{cases}$ be a curve which under the above transformation goes into the line L: $\alpha + \beta = 0$, $\alpha \leq \alpha_0$, $\beta \leq \beta_0$ where $\alpha_0 + \beta_0 > 0$. (So C is not a characteristic curve.)

(d) Let $z_0(\mu)$, $p_0(\mu)$, $q_0(\mu)$, $r = r_0(\mu)$, $s = s_0(\mu)$, $t = t_0(\mu)$ be given functions which can be transformed correspondingly into functions of arc length and which satisfy (24.4) (strip equations). Then there exists a function $\varphi(x,y)$ such that, for some region R containing C,

(α) $\varphi(x,y)$ is of class A^1 in R

(β) $z = \varphi(x,y)$ is a solution of (22.2) in R

(δ) $\varphi[x_0(\mu),y_0(\mu)] = z_0(\mu)$, $\varphi_x[x_0(\mu),u_0(\mu)] = p_0(\mu)$,

$$\varphi_y[x_0(\mu),y_0(\mu)] = q_0(\mu),\ \varphi_{xx}[x_0(\mu),y_0(\mu)] = r_0(\mu),$$

$$\varphi_{xy}[x_0(\mu),y_0(\mu)] = s_0(\mu),\ \varphi_{yy}[x_0(\mu),y_0(\mu)] = t_0(\mu).$$

Cinquini-Cibrario[4] solved problem C by reducing it to a problem in first order equations as above and then using Picard successive approximations. Cibrario[2,3] used this same method to solve the problem of Goursat (the second paper listed simpler hypotheses than the first).

THEOREM 37.2:

(a) Let $F(x,y,z,p,q,r,s,t)$ be of class A^3 and let

$\Delta = 4F_r F_t - F_s^2 < 0$ in a region S.

(b) Let γ_1 and γ_2 be two arcs given by equations of class A^3, intersecting at an interior point 0.

(c) Let $\sigma(\mu)$ and $\tau(\nu)$ be functions given on these arcs of class A^3, and let all functions $x(\mu), y(\mu)$, $\bar{x}(\nu), \bar{y}(\nu)$, $\sigma(\mu), \tau(\nu)$ satisfy Lipschitz conditions. Let γ_1 and γ_2 have distinct tangents at 0 which are not divided harmonically by the projections on the x-y plane of the characteristic directions at 0.

Then there exists a unique function $\varphi(x,y)$ such that:

(α) $\varphi(x,y)$ is of class A^2 in a neighborhood N of 0.

(β) $z = \varphi(x,y)$ is a solution of (22.2) in N.

(γ) $\varphi(x(\mu),y(\mu)) = \sigma(\mu)$ and $\varphi(\bar{x}(\nu),\bar{y}(\nu)) = \tau(\nu)$.

COROLLARY 1: If (22.2) is the quasi-linear equation (22.1), A^3 may be replaced by A^2 in (b).

COROLLARY 2: If γ_1 and γ_2 are arcs having 0 as a common end point, same result holds in the angle between γ_1 and γ_2 provided this angle is free of projections formed by characteristic directions.

Cinquini-Cibrario[4] also considered the problem where φ and φ_x are given on γ_1 and just φ on γ_2. She proved the existence of a unique solution, provided α_1 and α_2 are separated by exactly one characteristic direction. The same result was found by Schauder.[1]

E. E. Levi[1] also considered equation (22.2) in his work on hyperbolic equations in n independent variables.

Both Hadamard[2] pp. 501-513 and Courant-Hilbert; pp. 332-337 have expositions of Lewy's theorem, in which the proofs are probably easier to follow than in the original paper. (C.-H. reduce the problem to quasi-linear equations, by successive approximation, which they then solve by Lewy's method pp. 326-332).

E. Systems of Hyperbolic Equations in Two Variables

38. Linear Systems

Germay[3] established the following results:

THEOREM 38.1: There exists a unique system of functions $z_1 = U(x,y)$ and $z_2 = V(x,y)$ which are analytic, satisfy the equations:

$$(38.1)\quad \frac{\partial^2 z_i}{\partial x \partial y} = a_i(x,y)\frac{\partial z_1}{\partial x} + b_i(x,y)\frac{\partial z_1}{\partial y} + c_i(x,y)z_1$$

$$+ g_i(x,y)\frac{\partial z_2}{\partial x} + h_i(x,y)\frac{\partial z_2}{\partial y} + k_i(x,y)z_2 + f_i(x,y)$$

$$(i = 1,2)$$

where $a_i, \ldots, f_i$ are continuous in D: $\begin{cases} 0 \le x \le \alpha \\ 0 \le y \le \beta \end{cases}$ (α, β finite positive numbers) and such that

$$U(x,0) = \varphi_1(x); \qquad U(0,y) = \psi_1(y)$$

$$V(x,0) = \varphi_2(x); \qquad V(0,y) = \psi_2(y)$$

where $\varphi_i(x)$ are continuous on $0 \le x \le \alpha$, $\psi_i(y)$ are continuous on $0 \le y \le \beta$, and

$\varphi_i(0) = \psi_i(0) \qquad (i = 1,2).$

Proof: (An extension of the Riemann method, using Green's functions.)

The above equations are a special case ($\lambda = 1$) of:

$$(38.2) \qquad \frac{\partial^2 z_i}{\partial x \partial y} = \lambda[a_i(x,y)\frac{\partial z_i}{\partial x} + \dots + k_i(x,y)z_2] + f_i(x,y)$$

Then the functions $U(x,y;\lambda)$ and $V(x,y;\lambda)$ whose existence is established in the above theorem are given by:

$$U(x,y;\lambda) = \int_0^x \int_0^y (\ f_1(\zeta,\eta)\ H_1(x,y;\zeta,\eta\ \lambda) + f_2(\zeta,\eta) \cdot H_2(x,y;\zeta,\eta;\lambda))\ d\eta d\zeta$$

$$V(x,y;\lambda) = \int_0^x \int_0^y (\ f_1(\zeta,\eta)\ K_1(x,y;\zeta,\eta;\lambda) + f_2(\zeta,\eta) \cdot K_2(x,y;\zeta,\eta;\lambda))\ d\eta d\zeta$$

where H_1, H_2, K_1, K_2 are the *Riemann functions* associated with 38.2.

These functions are analogous to Green's functions and are solutions of:

$$\frac{\partial^2 z_i}{\partial x \partial y} = [a_i(x,y)\frac{\partial z_1}{\partial x} + \dots + k_i(x,y)z_2] \qquad (i = 1,2),$$

one set being $\begin{cases} z_1 = H_1(x,y) \\ z_2 = K_1(x,y) \end{cases}$ and the other set being $\begin{cases} z_1 = H_2(x,y) \\ z_2 = K_2(x,y) \end{cases}$

for $x = \zeta$, they take on initial values which are solutions of:

$$\frac{d\rho}{dy} = \lambda[a_1(\zeta,y)\ \rho + g_1(\zeta,y)\sigma]$$

$$\frac{d\sigma}{dy} = \lambda[a_2(\zeta,y)\ \rho + g_2\ \zeta,y)\sigma]$$

and for $y = \eta$, they take on initial values which are solutions of:

$$\frac{d\theta}{dx} = \lambda[b_1(x,\eta)\theta + h_1(x,\eta)\omega]$$

$$\frac{d\omega}{dx} = \lambda[b_2(x,\eta)\theta + h_2(x,\eta)\omega].$$

The determinant $\begin{vmatrix} H_1 & H_2 \\ K_1 & K_2 \end{vmatrix} = 1$ for $x = \zeta$, $y = \eta$.

Results can be generalized, according to Germay, to systems of equations of the form:

$$(38.3) \quad \frac{\partial^2 z_i}{\partial x \partial y} = \lambda \sum_{k=1}^{p} [a_{jk}(x,y) \frac{\partial z_k}{\partial x} + b_{jk}(x,y) \frac{\partial z_k}{\partial y} + c_{jk}(x,y)\, z_k(x,y)] + f_j(x,y) \quad (j = 1,\dots,p)$$

to obtain solutions which vanish for $x = 0$ and $y = 0$.

Solutions are $z_j = z_j(x,y;\lambda) = \sum_{s=1}^{p} \int_0^x\int_0^y f_s(\zeta,\eta)\, G_{js}(x,y;\zeta,\eta;\lambda)\, d\eta d\zeta \quad (j = 1,\dots,p)$

where the p^2 *Riemann functions* G_{js} are defined analogously to the above and are given by:

$$G_{js}(x,y;\zeta,\eta;\lambda) = \sum_{n=0}^{\infty} \lambda^n g_{js}^{(n)}(x,y;\zeta,\eta) \quad (j,s = 1,\dots,p)$$

where
$$g_{js}^{(n)}(x,y;\zeta\,\eta) = \int_\zeta^x\int_\eta^y \sum_{k=1}^{p} [a_{jk}(u,v) \frac{\partial g_{ks}^{(n-1)}(u,v;\zeta,\eta)}{\partial u} + b_{jk}(u,v) \frac{\partial g^{(n-1)}}{\partial v} + c_{js} g_{ks}^{(n-1)}]\, dvdu$$
$$+ \int_\eta^y \sum_{k=1}^{p} [a_{jk}(\zeta,v) g_{ks}^{(n-1)}(\zeta,v;\zeta,\eta)]\, dv$$
$$+ \int_\zeta^x \sum_{k=1}^{p} [b_{jk}(u,\eta) g_{ks}^{(n-1)}(u,\eta,\zeta,\eta)\, du$$
$$(j = 1,\dots,p;\ n = 1,2,3,\dots)$$

and these recursion formulas begin with

$$g_{js}^{(0)} = 0 \text{ for } j \neq s; \quad g_{ss}^{(0)} = 1.$$

39. For quasi-linear systems:

$$\frac{\partial^2 z_i}{\partial x \partial y} = a_i(x,y)\frac{\partial z_1}{\partial x} + b_i(x,y)\frac{\partial z_1}{\partial y} + c_i(x,y)z_1$$

$$(39.1) \qquad + g_i(x,y)\frac{\partial z_2}{\partial x} + h_i(x,y)\frac{\partial z_2}{\partial y} + k_i(x,y)z_2$$

$$+ F_i(x,y;z_1.z_2;\frac{\partial z_1}{\partial x}, \frac{\partial z_1}{\partial y}, \frac{\partial z_2}{\partial x}, \frac{\partial z_2}{\partial y}) \qquad (i = 1,2)$$

Germay[3] **showed** that, in the neighborhood of (0,0), there exist solutions

$$(39.2) \qquad z_1 = U(x,y), \quad z_2 = V(x,y)$$

of (39.1) such that (39.3) $U(x,0) = U(0,y) = V(x,0) = V(0,y) = 0$

provided $a_i, \ldots, k_i$, $F_i(x,y;z_1,z_2,p_1,q_1,p_2,q_2)$ are continuous in

$$D: \; (0 \leq x \leq \alpha, \; 0 \leq y \leq \beta, \; |z_i| \leq R_i; \; |p_i| \leq P_i; \; |q_i| \leq Q_i$$

$$(i = 1,2) \text{ in } D.$$

and provided F_i satisfy Lipschitz **conditions**:

$$|F_i(x,y;z_1,z_2,p_1,q_1,p_2,q_2) - F_i(x,y;\bar{z}_1,\bar{z}_2,\bar{p}_1,\bar{q}_1,\bar{p}_2 \; \bar{q}_2)|$$

$$\leq M_i[\,|z_1-\bar{z}_1| + |z_2-\bar{z}_2| + \ldots + |q_2-\bar{q}_2|\,] \qquad (i = 1,2)$$

The procedure is a recursive one:

$$z_1 = U_0(x,y), \quad z_2 = V_0(x,y)$$

are solutions of the linear system obtained from (39.1) by neglecting the last term, which vanish for $x = 0$ and $y = 0$.

$z_1 = U_{n+1}(x,y)$ and $z_2 = V_{n+1}(x,y)$ are solutions of the linear system:

$$\frac{\partial^2 z_i}{\partial x \partial y} = a_i(x,y)\frac{\partial z_1}{\partial x} + \ldots + k_i(x,y)z_2 +$$

$$F_i[x,y;U_n,V_n;\frac{\partial U_n}{\partial x}, \frac{\partial U_n}{\partial y}; \frac{\partial V_n}{\partial x}, \frac{\partial V_n}{\partial y}] \qquad (i = 1,2)$$

which vanish for $x = 0$ and for $y = 0$. Then the solution of (39.1) and (39.2) is $U = \lim_{n\to\infty} U_n(x,y)$ and $V = \lim_{n\to\infty} V_n(x,y)$.

He also considers the general systems

$$(39.2) \qquad \frac{\partial^2 z_i}{\partial x \partial y} = \Phi_i(x,y;z_1,z_2;\frac{\partial z_1}{\partial x},\frac{\partial z_1}{\partial y};\frac{\partial z_2}{\partial x},\frac{\partial z_2}{\partial y}) \qquad (i = 1,2)$$

where $\Phi_i(x,y;z_1,z_2;p_1,q_1;p_2,q_2)$ are of class A^{∞} in the neighborhood of (0,0,0,0,0,0,0,0) and such that the linear part of their expansions in powers of $z_1 z_2 p_1 p_2 q_1 q_2$ do not vanish identically. Then the above system has a unique solution:

$$z_1 = U(x,y), \quad z_2 = V(x,y)$$

such that $U(x,y)$ and $V(x,y)$ are of class A^{∞} in a neighborhood of (0,0) and

$$U(0,y) = V(0,y) = U(x,0) = V(x,0) = 0.$$

These integrals can be calculated by successive approximation by means of Riemann functions associated with the linear parts of the expansion of the second members, for $\lambda = 1$.

Cioranescu[1] considered solutions of 39.1 and 39.2 by solving equivalent systems of first order equations. L. Martin[1] also studied (39.1), getting solution of Cauchy problem.

Kourensky, M.[1] in a series of papers obtained necessary conditions that systems of the following form be compatible:

$$\begin{cases} F_1(x,y,z,\bar{z},p,q,\bar{p},\bar{q},r,s,t,\bar{r},\bar{s},\bar{t}) = 0 \\ F_2(x,y,z,\bar{z},p,q,\bar{p},\bar{q},r,s,t,\bar{r},\bar{s},\bar{t}) = 0 \end{cases}$$

in which certain specified partial derivatives of second order appear, are compatible with

$$\Phi\,(x,y,z,\bar{z},p,q,\bar{p},\bar{q},r,s,t,\bar{r},\bar{s},\bar{t}) = \text{constant}.$$

E. Titt[1], following the work of Riquier and Thomas, considered the completely integrable system:

$$(S)\quad \frac{\partial^2 v_i}{\partial x^r \partial x^s} = \sum_{h=1}^{w} \sum_{\alpha=1}^{s} \sum_{\beta=1}^{n} g_{irs}^{h\alpha\beta} \frac{\partial^2 v_h}{\partial x^\alpha \partial y^\beta} + \text{terms of lower order}$$

$$(i = 1,\ldots,L_{rs};\ r = 1,\ldots,s;\ s = 1,\ldots,n)$$

where no left member appears on right in any equation, and $g_{irs}^{h\alpha\beta} = 0$ if $s>\beta$ or if $s = \beta$ and $r>\alpha$. We say such a system is in regular form.

Given

$$\sum_{h=1}^{w} \sum_{\alpha,\beta=1}^{n} a_i^{h\alpha\beta} \frac{\partial^2 v_h}{\partial x^\alpha \partial x^\beta} + c_i = 0 \quad (i = 1,\ldots,L),$$

where $a_i^{h\alpha\beta}$ and c_i are functions of x^α, v_h, and $\frac{\partial v_h}{\partial x^\alpha}$, and L equations are linearly independent on $\frac{\partial^2 v_h}{\partial x^\alpha \partial x^\beta}$, then in a non-singular coordinate system, this system can always be put into regular form. Indeed if $a_i^{h\alpha\beta}$ and c_i are analytic functions of x^α, v_h and $p^{h\alpha}$ in neighborhood of $x^\alpha = 0,\ldots$ then the regular system is orthonomic.

THEOREM 39.3 There exists one and only one set of functions $v_{hab}(x)$, $(a,b \neq 0)$ analytic in neighborhood of $x^\alpha = 0$ such that together with first derivatives, they reduce to assigned functions on individual surfaces:

$$\begin{cases} v_{hab},\ \dfrac{\partial v_{hab}}{\partial x^j} \text{ reduce to } F(x^b,\ldots,x^n) \text{ on } x^1 = 0,\ldots,x^{b-1} = 0 \\ \qquad (j = a+1,\ldots,b-1) \\ v_{hab},\ \dfrac{\partial v_{hab}}{\partial x^m} \text{ reduce to } F(x^{b+1},\ldots,x^n) \text{ on } x^1 = 0,\ldots,x^b = 0 \\ \qquad (m = 1,\ldots,a) \end{cases}$$

and such that together with $v_{koo}(x)$ they constitute a solution of the completely integrable system (S).

The rest of Titt's paper is concerned with getting characteristic surfaces, and similar problems.

F. Hyperbolic Equations in More than Two Variables

40. In section 26, a classification was given of equations of the form:

$$(40.0)\qquad \sum_{i,k=1}^{n} a_{ik}(x_1,\dots,x_n)\frac{\partial^2 u}{\partial x_i\partial x_k} + f(x_1,\dots,x_n,u,\frac{\partial u}{\partial x_i},\dots\frac{\partial u}{\partial x_n}) = 0$$

If equation (40.0) is hyperbolic, the problem of Cauchy would be to determine a solution so that on a given hypersurface: $x_i = x_i^{\,0}(s_1,\dots,s_{n-1})$, we would have

$$U = u^0(s_1,\dots,s_{n-1}) \text{ and } u_{x_i} = p_i^{\,0}(s_1,\dots,s_{n-1}) \qquad (i=1,\dots,n)$$

where $x_i^{\,0}, u^0, p_i^{\,0}$ are arbitrary functions of class A^1 and

$$\frac{\partial u^0}{\partial s_j} = \sum_{i=1}^{n} p_j^{\,0}\,\frac{\partial x_i^{\,0}}{\partial s_j}.$$

If the surface were given in the form: $G(x_1,\dots,x_n) = 0$, then the initial data would be the value of u and of one of its directional derivatives, in any direction not tangent to the surface. Now if one wished to apply the Cauchy-Kowalewski theorem of section 56, it would be necessary to make a transformation such that this surface goes into a plane - say $X_n = 0$ So we can set $X_n = G(x_1,\dots,x_n)$, $X_i = x_i$ $(i=1,\dots,n-1)$. But then the resulting equation would need to be solved for $\frac{\partial^2 u}{\partial X_m^2}$, in order to bring it into normal form, which would mean that

$$\sum a_{ik}\,\frac{\partial G}{\partial x_i}\frac{\partial G}{\partial x_k} \neq 0$$

In that case, the solution would be unique. If the condition does not hold, we can say that the surface is a characteristic surface, which is exactly the definition given in section 26. In 1923, and in a revised form in 1932, Hadamard[1] made a comprehensive study of the problem of Cauchy for linear equations:(26.1) which we shall summarize. Defining bicharacteristics and conoids as in section 26, the geodesic distance $\Gamma(x_1, \ldots, X_n;\ a_1, \ldots, a_n)$ defined there may be obtained from:

$$\Sigma\, a_{ik} \frac{\partial\Gamma}{\partial x_i} \frac{\partial\Gamma}{\partial x_k} = 4\Gamma$$

He then establishes the following:

THEOREM 40.1: Consider the linear homogeneous equation:

$$(40.1)\quad L(u) \equiv \sum_{i,k=1}^{n} a_{ik} \frac{\partial^2 u}{\partial x_i \partial x_k} + \sum_{i=1}^{n} b_i \frac{\partial u}{\partial x_i} + cu = 0$$

where a_{ik}, b_i, c are functions of $(x_1, \ldots, x_n)$ and where the characteristic form has $|\bar{a}_{ik}| \neq 0$. Then there is an elementary solution with any arbitrary point $a = (a_1, \ldots, a_n)$ as pole. It is of the form:

$$(40.2)\quad u = \frac{U}{\Gamma^{\frac{m-2}{2}}} \text{ (m odd)}; \qquad u = \frac{U}{\Gamma^{\frac{n-2}{2}}} + V\log_e\Gamma \text{ (n even)}$$

where U is a specific holomorphic function taking on value $\frac{1}{\sqrt{|\Delta|}}$ at a, while V is any other holomorphic function. Thus it is unique, for odd n, and not unique for even n.

This fundamental result has been obtained in other and simpler ways, recently. Thomas and Titt[1] give it in terms of modern tensor analysis; Bureau[1] rephrases it and proves it by tensor notation also; Geheniau and De Donder have a form which they say is good for certain applications.

To solve the problem of Cauchy, first find the elementary solution of the adjoint equation:

$$(40.3)\quad M(v) = \sum_{i,k=1}^{n} \frac{\partial^2}{\partial x_i \partial x_k}(A_{ik}v) - \sum_{i=1}^{n} \frac{\partial}{\partial x_i}(B_i v) + Cv = 0.$$

The following identity relates L(u) and M(v); it is similar to that for n=2.

$$(40.4)\quad \iiint [v\, L(u) - u\, M(v)]\, dx_1, .. dx_n = -\iint_T \left(v \frac{\partial u}{\partial v} - u \frac{dv}{dv} + Nuv\right) dS$$

For the normal hyperbolic equation - that is, the one where the characteristic form $\Sigma\, a_{iK}\, \alpha_i\, \alpha_K$ can be reduced to $\beta_1^2 - \sum_{i=2}^{n} \beta_i^{\,2}$ at any given point, of a region R, Hadamard then obtains a solution of Cauchy's problem provided the characteristic conoid through a given point cuts the initial surface in a certain portion which together with the conoid will be a closed region - the formulas are too long to be given here, but they are analogous to the case n=2 and follow from substituting in (40.4). The solution is obtained for n odd, where it is unique, and then by a method of descent, for n even, where it is not unique, because of the last theorem. (His distinction of " space-oriented" and " time-oriented" surfaces is good when one of the variables associated with (40.1) is t.) Bureau[1] gave another kind of integral to replace the peculiar improper integral Hadamard had to use. The same is true of Thomas and Titt's[1] work, since they used normal coordinates. Schauder[1] considered the initial value problem for the linear hyperbolic equation in n variables:

$$(40.5)\quad \sum_{i,k=1}^{n} a_{ik} \frac{\partial^2 z}{\partial x_i \partial x_k} + \sum_{j=1}^{n}{}' b_j \frac{\partial z}{\partial x_j} + cz = F$$

where each a_{ik}, b_i, c, F is of class A^{∞} in the neighborhood of every point $(x_1^{\,0}, \ldots, x_n^{\,0})$.

THEOREM 40.2: There exists a number $\delta > 0$ depending only on the coefficients and on the *pyramid frustrum* P_n so that for any arbitrary plane E: x_n = constant and for arbitrary initially analytic data on E, the Cauchy problem for (40.5) is solved on every strip of E of width δ. Here P_n is supposed to be pyramid with base in an (n-1) dimensional cube in the plane $x_n = x_n^0$.

Then by replacing condition of analyticity by condition that coefficients and data can be approximated by analytic data, Schauder widens the existence region of the solution to be an entire pyramid $P_n(M_1)$ where $|ao_k| \le M_1$, $|b_i| \le M_1$, $|c| \le M_1$, $|da_{ik}| \le M_1$. Lahaye[1] obtained a solution to the problem of Cauchy for equations of the form:

$$\left\{A_0(x) + \sum_{i=1}^{n} A_i(x) \frac{\partial}{\partial x_i}\right\} \left\{\sum_{i=1}^{n} \beta_i(x) \frac{\partial u}{\partial x}\right\} = E(x) \qquad (x=(x_1,..x_n))$$

and then applied successive approximation to get solution of the problem for (40.1).

Mathisson[1] obtained a solution to (40.5) for any hyperbolic equation by using a Reimann metric and covariant differentiation.

41. Special Linear Equations

Consider the problem of finding a solution of

$$(41.1) \quad \frac{\partial^2 z}{\partial t^2} - \frac{\partial^2 z}{\partial x^2} - \frac{\partial^2 z}{\partial y^2} = 0$$

of the form $z = \varphi(t,x,y)$ so that

$$(41.2) \qquad \varphi(0,x,y) \equiv z_0(x,y)$$

$$\varphi_1(0,x,y) \equiv q_0(x,y)$$

We set $z_{0,x} = \frac{\partial z_0}{\partial x}(x,y)$ and $z_{0,y} = \frac{\partial z_0}{\partial y}(x,y)$ and let $\Gamma \equiv t^2 - (X-x)^2 - (Y-y)^2$.

The Volterra formula is:

$$z(t,x,y) = \frac{1}{2\pi t} \iint_R \frac{1}{\sqrt{\Gamma}} \quad z_0(X,Y) + t\, q_0(X,Y) + (X-x) z_{0,x}(X,Y)$$

$$+ (Y-y) z_{0,y}(X,Y) \qquad dXdY$$

where R is the region in the (X,Y) plane defined by $\Gamma \ge 0$ - that is, the interior and boundary of the circle: $(X-x)^2 + (Y-y)^2 = t^2$. One can also write

$$z(t,x,y) = \frac{1}{2\pi}\int_0^{2\pi}\int_0^1 \frac{1}{\sqrt{1-\rho^2}} \left\{ z_0 + tq_0 + t\rho \cos \theta z_{0,x} + t\rho \sin \theta z_{0,y} \right\} \rho d\rho d\theta$$

where arguments of z_0, q_0, z_{0x}, z_{0y} are $(x+T\rho \cos \theta,\ y+t\rho \sin \theta)$. Existence of solutions is assured provided z_0 is of class A^3 and q_0 of class A^2; then $\varphi(t,x,y)$ will be of class A^2. Above formulas show that if z_0 is of class A^{n+1} and q_0 of class A^n, then we can get formulas for nth derivatives of φ, and so φ is of class A^n.

The *Huygens principle in the extended sense* says that the solution on the plane $t = t_2$ $(t_2>0)$ instead of being obtained directly for $t = 0$, can also be constructed by first finding solution on $t = t_1$ $(0<t_1<t_2)$ with initial conditions on $t = 0$ and then finding solution on $t = t_2$ with new initial conditions on $t = t_1$. (Hadamard[1]) This will require if we want $z = \varphi(t,x,y)$ to be of class A^2, that values on $t = t_2$ are of class A^3 and on $t = 0$ of class A^4.

Courant and Lewy[3] obtain existence of solutions of (41.1) and (41.2) using another kind of condition. Suppose E is a convex closed region in the (ζ,η) plane. Denote by $\Delta^n Z$ any partial derivative:

$$\frac{\partial^n Z}{\partial x^k \partial y^{n-k}} \qquad (h=0,\ldots,n).$$

A continuous function $w(\zeta,\eta)$ defined in E has the property A provided:

A) $w(\zeta,\eta)$ can be uniformly approximated in E by functions $w_\nu(\zeta,\eta)$ of class A^2 such that

$$\iint_E [\Delta w_\nu]^2 \, d\zeta d\eta \leq M \quad \text{and} \quad \iint_E [\Delta^2 w_\nu]^2 \, d\zeta d\eta \leq N \quad (\nu=1,2,3\ldots)$$

hold for all possible Δw_ν and $\Delta^2 w_\nu$ and for certain $M > 0$, $N > 0$ which are free of ν.

A function $v(\zeta,\eta)$ defined in E has the property A_n provided:

A_n) $v(\zeta,\eta)$ is of class A^n in E; v and its derivatives up to nth order can be uniformly approximated in E by functions $v_\nu(x,y)$ of class A^{n+2} where the nth derivatives of v_ν have property A; that is,

$$\iint_E [\Delta^{n+1} v_\nu]^2 \, d\zeta d\eta \le M \quad \text{and} \quad \iint_E [\Delta^{n+2} v_\nu]^2 \, d\zeta d\eta \le N$$
$$(\nu = 1,2,3\ldots)$$

for all possible $\Delta^{n+1}v_\nu$ and $\Delta^{n+2}v_\nu$, and certain $M > 0$, $N > 0$ which are free of ν.

LEMMA: If v has property A_n then $v_\nu(x,y)$ can be taken as polynomials. (Weierstrass approximation.)

THEOREM 41.1 If $z_0(x,y)$ and $q_0(x,y)$ are polynomials there exists a polynomial $\varphi(t,x,y)$ satisfying (41.1) and (41.2). (This is immediate.)

THEOREM 41.2 Let $z_0(x,y)$ and $q_0(x,y)$ be defined in E and posses property A_2 and property A_1 respectively. Let K be a frustrum of a cone with base in the circle: $x^2 + y^2 \le R^2$, in the plane $t = 0$ and vertex ($t = R > 0$; $x = y = 0$), cut off by plane $t =$ constant.

Then there exists a function $\varphi(t,x,y)$ defined in K which is of class A^2, which has the property A_2 in every plane P whose distance from the x-y plane is $< \pi_4$ and such that $\frac{\partial\varphi}{\partial n}$ has property A_1 in P ($\frac{\partial\varphi}{\partial n}$ is in direction of normal to P) ; the function $\varphi(t,x,y)$ satisfies (41.1) and (41.2).

Michlin[1] considers problems of Cauchy, Dirichlet, and Neumann type for equation (41.1) on a half cylinder.

Courant, Friedrichs, and Lewy[1] consider what is essentially the same equation:

$$2\frac{\partial^2 z}{\partial t^2} - \frac{\partial^2 z}{\partial x^2} - \frac{\partial^2 z}{\partial y^2} = 0$$

by means of difference equations. Existence is established provided z_0 and q_0 are of class A^4. Solution is of class A^2. In the same paper they consider the more general equation:

$$\frac{\partial^2 z}{\partial t^2} - \left(a\frac{\partial^2 z}{\partial x^2} + 2b\frac{\partial^2 z}{\partial x\partial y} + c\frac{\partial^2 z}{\partial y^2}\right) + \alpha\frac{\partial z}{\partial t} + \beta\frac{\partial z}{\partial x} + \gamma\frac{\partial z}{\partial y} + \delta z = 0$$

with boundary conditions (41.2), where $a > 0$, $c > 0$, $ac - b^2 > 0$ and all coefficients are of class A^3, while z_0 is of class A^4 and q_0 is of class A^3. Solution is again obtained by using difference equations.

Ignatovskij[1] considers the same kind of problem in four variables:

$$\frac{1}{c^2}\frac{\partial^2 u}{\partial t^2} - \left(\frac{\partial^2 u}{\partial x^2} + \frac{\partial^2 u}{\partial y^2} + \frac{\partial^2 u}{\partial z^2}\right) + c_1\frac{\partial u}{\partial t} + mu = \varphi(x,y,z,t).$$

Using Laplace transforms, he solves problems of special kinds for this equation (no proofs). For the case $c_1 = m_1 = \varphi = 0$, see Feller and Tamarkin[1], p. 47-49, for treatment by Hadamard's methods.

Lowan[1,2] has considered a series of problems with the equation:

$$\nabla^2 u = 2pu_t + \frac{1}{a^2}u_{tt} + \phi$$

$$\nabla^2 = \frac{\partial^2}{\partial x^2} \text{ or } \frac{\partial^2}{\partial x^2} + \frac{\partial^2}{\partial y^2} \text{ or } \frac{\partial^2}{\partial x^2} + \frac{\partial^2}{\partial y^2} + \frac{\partial^2}{\partial z^2}\ ;\ \phi = \phi\,)P,t),$$

P being either x or (x,y) or (x,y,z).

Boundary conditions are of the nature of problem G or problem I.

The solution is obtained by Laplace transforms. This method, of reducing the number of variables involved by a Laplace transform, has been used by others.

Gogoladze[1] considered equations of the form:

$$A_1 \frac{\partial^2 u}{\partial x^2} + A_2 \frac{\partial^2 u}{\partial y^2} + A_3 \frac{\partial^2 u}{\partial z^2} + a_1 \frac{\partial u}{\partial x} + a_2 \frac{\partial u}{\partial y} + a_3 \frac{\partial u}{\partial z} + a_4 u = \frac{1}{c^2} \frac{\partial^2 u}{\partial t^2}$$

where A_i, a_i, c are functions of (x,y,z). Boundary conditions give Cauchy data u and u_t for $t = 0$ and value of u on a surface. He gets a generalized potential. Other special equations have been considered - the above selection indicates the methods used.

42. Quasi-Linear and General Case

Soboleff[1], using some inequalities generalized from a lemma of Friedrichs and Lewy, shows that there exists a solution $z = \varphi(x,y)$ of

$$\sum_{i,k=1}^{n} A_{ik} \frac{\partial^2 z}{\partial x_i \partial x_k} - \frac{\partial^2 z}{\partial t^2} = F$$

where A_{ik}, F depend on

$$z, \frac{\partial z}{\partial x_1}, \ldots, \frac{\partial z}{\partial x_n}, \frac{\partial z}{\partial t}, x_1, \ldots, x_n, t \text{ and } \sum_{i,k=1}^{n} A_{ik} p_i p_k > C \sum_{i=1}^{n} p_i^2$$

satisfying the conditions:

$$u(x,0) = z_0(x), \quad u_t(x,0) = q_0(x) \qquad x = (x_1, \ldots, x_n)$$

and such that φ is of class A^2. The solution is unique. Schauder obtained essentially the same result on the basis of polynomial approximations together with his results on linear equations. Kryzanski and Schauder[1] considered mixed boundary problems for this equation.

Titt[1] treats a fairly general analytic non-linear problem with various kinds of initial data by reducing it to linear problems, by Lewy's method, when $n = 3$.

G. Parabolic Equations

43. The equation

$$(43.1) \quad a(x,y)\frac{\partial^2 z}{\partial x^2} + 2b(x,y)\frac{\partial^2 z}{\partial x \partial y} + c(x,y)\frac{\partial^2 z}{\partial y^2} + f(x,y,z,\frac{\partial z}{\partial x},\frac{\partial z}{\partial y}) = 0$$

$$(ac = b^2),$$

can be reduced to a canonical form:

$$(43.1a) \quad C(\zeta,\eta)\frac{\partial^2 \omega}{\partial \zeta^2} + F(\zeta,\eta,\omega,\frac{\partial \omega}{\partial \zeta},\frac{\partial \omega}{\partial \eta}) = 0 \qquad (C \neq 0)$$

by a transformation T: $\zeta = U(x,y)$, $\eta = V(x,y)$ provided

$$(43.2) \quad |a| + |c| > 0$$

in the region P_0. (Shown on page 102.) There is only one family of characteristics of (43.1), since the characteristic equation (25.1) becomes

$$a \cdot y'^2 - 2\sqrt{ac}\; x'y' + c \cdot x'^2 = 0$$

(assuming $a \geq 0$, $c \geq 0$ since $ac = b^2 \geq 0$). For the normal form,

$$(43.3) \quad \frac{\partial^2 z}{\partial x^2} + f(x,y,z,\frac{\partial z}{\partial x}, \frac{\partial z}{\partial y}) = 0$$

these are the lines:

$$(43.4) \quad y = k$$

If a, b, c, are constants, A, B, C, with $AC = B^2$, because of (43.2) $A \neq 0$ and $C \neq 0$, and both have same sign so we take $A > 0$, $C > 0$. The characteristics are the lines: $\sqrt{A}\,y - \sqrt{C}\,x = k$; the characteristic through (x_0, y_0) is:

$$\sqrt{A}\,(y-y_0) = \sqrt{C}\,(x-x_0).$$

The transformation:

$$(43.5) \quad \xi = \sqrt{A}\,x + \sqrt{C}\,y;$$

$$\eta = -\sqrt{C}\,x + \sqrt{A}\,y$$

will reduce the equation to the form (43.1a).

44. There are three basic types of problems for parabolic equations. The problem of Cauchy is to find a solution such that z and its normal derivative assume prescribed values along a curve which is nowhere tangent to a characteristic. For the normal form (43.3), the initial value problem I ix to find a solution such that $\varphi(x_0, y) = g(y)$ and $\varphi_x(x_0, y) = h(y)$. (Problem C can be reduced to this, at least in the neighborhood of a point.) For analytic data, the existence of such a solution is given by theorem 25.1. When the initial data is not analytic, Holmgren[1] showed that there is not necessarily a solution to the problem; this is true even if we require only that:

$$(44.1) \quad \lim_{x \to x_0^+} \varphi(x_0, y) = g(y), \quad \lim_{x \to x_0^+} \varphi_x(x, y) = h(y).$$

The second kind of problem, which we shall call problem H(sometimes called the problem of limits) in general consists in finding a solution which assumes prescribed values along a portion of a characteristic, and along two adjacent non-characteristic curves. E. E. Levi[1] gave a classification of such frontier curves. One or both of these adjacent curves may be a line at infinity.

The third type of problem, problem D, consists in finding a solution which assumes prescribed values along a closed contour; this is analogous to the Dirichlet problem for elliptic equations, but it does not always have a solution, as was shown by Gevrey[1].

(Doetsch[1], in his address before the International Conference on Partial Differential Equations held in Geneva in 1936, gives an excellent resume of the historical development of the idea of a solution - including why (44.1) needs to be used. He discusses many other things besides existence theorems, such as uniqueness.)

45. In this section we consider the simplest parabolic equation. Holmgren[1]

considered:

$$(45.1) \quad \frac{\partial^2 z}{\partial x^2} - \frac{\partial z}{\partial y} = 0$$

A necessary and sufficient condition that there exist a solution of (45.1) for $a \leq y \leq b$, $x_0 \leq x$ which is regular and for which (44.1) holds, where $g(y)$ is of class A^1 is that

$$h(y) + \frac{1}{\sqrt{\pi}} \int_a^y \frac{g'(\eta)}{\sqrt{y-\eta}} \, d\eta$$

be a *function H of class* 2. This last term is due to Gevrey[1], who in his fundamental paper on parabolic equations, called a function $f(z)$ a function H of class α ($\alpha \geq 1$) if $f(z)$ has an infinite number of derivatives on an interval and if $|f^{(n)}(z)| \leq \frac{M(n!)^\alpha}{r^n}$.

If we consider a region R:

$$\begin{cases} \varphi_1(t) \leq x \leq \varphi_2(t) \\ t_1 \leq t \leq t_2 \end{cases}$$

where $\varphi_1'(t)$ and $\varphi_2'(t)$ are bounded, then Gevrey[1] showed in 1913 that in order to find a solution $u = \varphi(x,t)$ of $u_t = u_{xx}$ such that it will take on given values on the boundary B, the functions $\varphi_1(t)$ and $\varphi_2(t)$ must satisfy *Holder conditions* with $\alpha > ½$. That is, for sufficiently small h,

$$|\varphi_i(t+h) - \varphi_i(t)| < C\,|h|^\alpha \qquad (i=1,2)$$

where α is a constant. Petrowsky[2] generalized this result as follows:

THEOREM 45.1: There will exist a solution of

$$(45.2) \quad \frac{\partial u}{\partial t} = \frac{\partial^2 u}{\partial x^2}$$

in R such that on the boundary of R, $u(x,t) = f(x,t)$, a continuous function, if for every t there exists a function $\rho(h)$ which is continuous, positive, defined for $h < 0$, has $\lim_{h \to 0^-} \rho(h) = 0$ and for $|h|$ sufficiently small,

$$(45.3) \qquad \begin{aligned} \varphi_1(t+h) - \varphi_1(t) &\geq 2\sqrt{h \log \rho(h)} \\ \varphi_2(t+h) - \varphi_2(t) &\leq 2\sqrt{h \log \rho(h)} \end{aligned}$$

and

$$\int_c^{\varepsilon} \frac{\rho h \sqrt{|\log \rho(h)|}\, dh}{h} \to -\infty \text{ as } h \to 0^- \quad (c < 0).$$

Conversely, there exists a function f such that there is no solution $u(x,t)$ of (45.2) in G assuming these values on B, in case for at least one t there is a function $\rho(h)$, continuous, positive, $\to 0$ as $h \to 0^-$ (defined for $h < 0$), and such that (45.3) holds, but

$$\int_c^{\varepsilon} \frac{\rho h \sqrt{|\log \rho(h)|}\, dh}{h}$$

converges for $\varepsilon \to 0^-$.

We now give more explicit solutions for certain problems of of type H. The characteristics are the lines t = constant.

The equation

$$(45.4) \qquad \frac{\partial u}{\partial t} = a^2 \frac{\partial^2 u}{\partial x^2}$$

where a is a constant, is the basic equation of heat conduction.

Problem H_1: To find a solution of (45.4) such that it is defined for

$$0 \leq x \leq L \quad , \quad t \geq 0 \quad ,$$

such that

$$u = 0 \text{ for } x = 0 \text{ and } x = L$$

and

$$u = f(x) \text{ for } t = 0$$

If we write $u = T(t)\ v(x)$, then equation (45.4) breaks up into

$$\begin{cases} \dfrac{d^2v}{dx} + \lambda v = 0; \\ v = 0 \text{ at } x = 0 \text{ and } x = L \end{cases} \qquad \begin{cases} \dfrac{dT}{dt} + a^2\lambda T = 0 \\ T(0)v(x) = f(x). \end{cases}$$

For the first equation, when $\lambda = \frac{n^2\pi^2}{L^2}$, $(n=1,2,3\ldots)$ we get $v = \sin \frac{n\pi x}{L}$.

The second equation, for corresponding values of $\lambda = \lambda_n$ has solution:

$$T = T_n = C_n e^{-a_n\lambda_n t} \qquad (\lambda_n = \frac{n^2\pi^2}{L^2}) \qquad (C_n \text{ arbitrary})$$

Hence $u_n = C_n e^{-a_n \frac{n^2\pi^2}{L^2} t} \sin \frac{n\pi x}{L} \quad (n=1,2,\ldots)$

satisfy first two relations; in general, u_n will not satisfy $T(0)v(x) = f(x)$, since this would mean $C_n \sin \frac{n\pi x}{L} = f(x)$. However, if we write

$$(45.5) \qquad u(x,t) = \Sigma\, C_n e^{-a_n \frac{n^2\pi^2}{L^2} t} \sin \frac{n\pi x}{L}$$

then this condition would mean

$$f(x) = \sum_1^\infty C_n \sin \frac{n\pi x}{L}$$

and by Fourier series, if $f(x)$ is piecewise continuous on $0 \leq x \leq L$, this is true, provided

$$(45.6) \qquad C_n = \frac{2}{L} \int_0^L f(x) \sin \frac{n\pi x}{L}\, dx.$$

One can show that each $|C_n| \leq C$, and hence the series in (45.5) converges uniformly. Thus:

THEOREM 45.2: The function $u(x,t)$ given by (45.5) and (45.6) is the solution to problem H_1, provided $f(x)$ is piecewise continuous on $0 \leq x \leq L$.

(This theorem can also be proved by using the Laplace transform. See Churchill[1].)

Problem H_2: To find a solution of (45.4) defined for R_0 $\begin{cases} -\infty<x<+\infty \\ t\geq 0 \end{cases}$ which is continuous in R_0 and such that

$$u = f(x) \text{ for } t = 0.$$

There is another condition here, sometimes not explicitly stated, but evident from the physical picture:

$$(45.7) \qquad \lim_{x\to\infty} u(x,t) = \lim_{x\to -\infty} u(x,t) = 0$$

It is not hard to find solutions of (45.4) which hold inside R, but do not exist for t = 0. Such a solution, satisfying (45.7) is:

$$(45.7a) \qquad \frac{1}{\sqrt{t}}\, e^{-\frac{x^2}{4a^2t}}$$

Hence problem H_2 is usually changed to:

Problem K: To find a solution $u = \varphi(x,t)$ of (45.4) such that $\varphi(x,t)$ is defined in R_1: $\begin{cases} -\infty<x<+\infty \\ t>0 \end{cases}$ is continuous in R_1 and such that

$$(45.8) \qquad \lim_{t\to 0^+} u(x,t) = f(x).$$

Then we have:

THEOREM 45.3: The function

$$u(x,t) = \frac{1}{2a\sqrt{\pi t}} \int_{-\infty}^{+\infty} f(\alpha)\, e^{-\frac{(\alpha-x)^2}{4a^2t}}\, d\alpha = \frac{1}{\sqrt{\pi}} \int_{-\infty}^{+\infty} f(x+2\beta a\sqrt{t})\, e^{-\beta^2}\, d\beta$$

satisfies (45.4) in R_1 and (45.8) holds

Proof: Formally it follows from (45.7a); to show it can be a solution involves some discussion of differentiation under the integral sign. (It can be obtained as a Fourier integral.) Courant, Friedrichs, Lewy[1] get it by means of a difference equation: $2\frac{(u_1 - u_0)}{L} = \frac{u_2 + u_4 - 2u_0}{h^2}$.

Other problems have been set up for (45.4) involving usually an *initial condition* like (45.8) and *boundary conditions* which are linear combinations of $u(0,t), u_x(0,t), u(L,t), u_x(L,t)$. Many have studied such problems by use of the Laplace transform. See, for example, Doetsch[1] and Churchill[1]. The same is true for the non-homogeneous equation:

$$(45.9) \quad \frac{\partial u}{\partial t} - \frac{\partial^2 u}{\partial x^2} = f(x,t)$$

The equation

$$h\frac{\partial u}{\partial t} = \frac{\partial}{\partial x}\left[p\frac{\partial u}{\partial x}\right] - qu \quad (h>0,\ p>0,\ q>0;\ \text{all three functions of } x)$$

has been studied; the problem is to determine a solution in

$$R_L \begin{cases} 0 \le x \le L \\ t \ge 0 \end{cases}$$

such that

$$\frac{\partial u}{\partial x} - hu = 0 \text{ for } x = 0$$

$$\frac{\partial u}{\partial x} + Hu = 0 \text{ for } x = L \qquad (h>0,\ H>0;\ h+H \text{ constants})$$

and

$$u = f(x) \text{ for } t = 0$$

Substituting $u = e^{-\lambda t} v$ as before, one obtains an ordinary differential equation, if $f(x)$ has properties which enable it to be expanded in a series:

$$f(x) = \sum_1^{\infty} c_n \varphi_n(x); \quad c_n = \int_0^L f(x)\, h(x)\, \varphi_n(x)\, dx$$

where $\varphi_n(x)$ are the normal functions associated with the *eigen values* λ of the problem:

$$\frac{d}{dx}\left(p\frac{dv}{dx}\right) + (h\lambda - q)\, v = 0$$

$$v'(0) - hv(0) = 0$$

$$v'(L) + Hv(L) = 0$$

then the solution of given problem is

$$u(x,t) = \sum_{n=1}^{\infty} c_n e^{-\lambda_n t} \varphi_n(x).$$

See Horn[1] - though the solution is found in many places.

46. *If $b \neq 0$ everywhere in R_0,* so that $ac > 0$ in R_0 and hence a and c maintain their signs, which can be taken as positive, the characteristic equation of (43.1) becomes:

$$\sqrt{a}\, y' - \sqrt{c}\, x' = 0$$

If $b \equiv 0$ in R_0, then $c \equiv 0$ and $a \neq 0$ everywhere in R_0 and characteristic equation becomes: $a\, x' = 0$ or $x' = 0$ and hence characteristics are (43.4), or else $a \equiv 0$ and $c \neq 0$ everywhere in R_0 and characteristics are x = constant.

In this section we give theorems applicable to the linear equation:

$$(46.1) \quad a(x,y)\frac{\partial^2 z}{\partial x^2} + 2b(x,y)\frac{\partial^2 z}{\partial x \partial y} + c(x,y)\frac{\partial^2 z}{\partial y^2} + d(x,y)\frac{\partial z}{\partial x} + e(x,y)\frac{\partial z}{\partial y} + g(x,y)z + h(x,y) = 0 \quad (ac = b^2)$$

where $a,\ldots,h$ are continuous in R_0: the reduced form (43.1a) is readily seen to be:

$$(46.2) \quad \frac{\partial^2 \omega}{\partial \zeta^2} + \bar{d}(\zeta,\eta)\frac{\partial \omega}{\partial \zeta} + \bar{e}(\zeta,\eta)\frac{\partial \omega}{\partial \eta} + \bar{g}(\zeta,\eta)\,\omega + \bar{h}(\zeta,\eta) = 0$$

where, setting $L(u) \equiv au_{xx} + 2bu_{xy} + cu_{yy} + du_x + eu_y$,

$$\bar{d}(\zeta,\eta) = \frac{L(U)}{[aU_x^2 + 2bU_xU_y + cU_y^2]}$$

$$\bar{e}(\zeta,\eta) = \frac{L(V)}{[aV_x^2+2bV_xV_y+cV_y^2]}$$

$$(46.3)\qquad \bar{g}(\zeta,\eta) = \frac{g(X,Y)}{[aU_x^2+2bU_xU_y+cU_y^2]} \qquad [x = X(\zeta,\eta);\ y = Y(\zeta,\eta)]$$

$$\bar{h}(\zeta,\eta) = \frac{h(X,Y)}{[aU_x^2+2bU_xU_y+cU_y^2]}$$

When a = A, b = B, c = C, where A, B, C are constants, then by (43.5), $u_{xx} = U_{xy} = U_{yy} = 0$ and similarly for V, so we get (46.2), with

$$\bar{d}(\zeta,\eta) = \frac{\sqrt{A}\,d(X,Y) + \sqrt{C}\,e(X,Y)}{(A+C)^2};\quad \bar{g}(\zeta,\eta) = \frac{g(X,Y)}{(A+C)^2}$$

$$\bar{e}(\zeta,\eta) = \frac{-\sqrt{C}\,d(X,Y) + \sqrt{A}\,e(X,Y)}{(A+C)^2};\quad \bar{h}(\zeta,\eta) = \frac{h(X,Y)}{(A+C)^2}$$

where $X(\zeta,\eta)$ and $Y(\zeta,\eta)$ are given by the inverse transformation:

$$T^{-1}\quad \begin{aligned} x &= X(\zeta,\eta) \equiv \frac{\sqrt{A}\,\zeta - \sqrt{C}\,\eta}{A+C} \\ y &= Y(\zeta,\eta) \equiv \frac{\sqrt{C}\,\zeta + \sqrt{A}\,\eta}{A+C} \end{aligned}$$

As a further special case of (46.1), we have d = D, e = E, g = G, all constants. The resulting equation:

$$(46.4)\qquad A\frac{\partial^2 z}{\partial x^2} + 2B\frac{\partial^2 z}{\partial x\partial y} + C\frac{\partial^2 z}{\partial y^2} + D\frac{\partial z}{\partial x} + E\frac{\partial z}{\partial y} + Gz + h(x,y) = 0$$

becomes under transformation (43.5)

$$(46.5)\qquad \frac{\partial^2 \omega}{\partial \zeta^2} + \bar{D}\frac{\partial \omega}{\partial \zeta} + \bar{E}\frac{\partial \omega}{\partial \eta} + G\omega + \bar{h}(\zeta,\eta) = 0$$

where $\bar{D} = \frac{\sqrt{A}\,D + \sqrt{C}\,E}{(A+C)^2}$, $\bar{E} = \frac{-\sqrt{C}\,D + \sqrt{A}\,E}{(A+C)^2}$ and $\bar{h}(\zeta,\eta)$ is defined as in (46.3).

But this can be simplified still further. Suppose, for example, $\bar{E} > 0$. Then by setting $S = \sqrt{-E}\xi$, (46.5) can be reduced to:

$$(*)\qquad \frac{\partial^2 \omega}{\partial s^2} - \frac{\bar{D}}{\sqrt{-E}}\frac{\partial \bar{\omega}}{\partial s} - \frac{\partial \bar{\omega}}{\partial \eta} - \frac{\bar{G}}{E}\,\bar{\omega} - \frac{\bar{h}}{E} = 0$$

and then by setting

$$v = \bar{\omega}\, e^{-\frac{1}{2}\frac{\bar{D}}{\sqrt{-\bar{E}}} s},$$

this reduces to:

$$(46.6) \quad \frac{\partial^2 v}{\partial s^2} - \frac{\partial v}{\partial \eta} + K v - \bar{h}(s,\eta) = 0.$$

When the equation is homogeneous so that h = 0, we get

$$\frac{\partial^2 v}{\partial s^2} - \frac{\partial v}{\partial \eta} + K v = 0,$$

and if $K \neq 0$, this can be further simplified by letting $u = e^{K\eta} v$, so that we get:

$$(46.7) \quad \frac{\partial^2 u}{\partial s^2} - \frac{\partial u}{\partial \eta} = 0.$$

In other words, the homogeneous linear equations with constant coefficients,

$$(46.8) \quad A \frac{\partial^2 z}{\partial x^2} + 2B \frac{\partial^2 z}{\partial x \partial y} + C \frac{\partial^2 z}{\partial y^2} + D \frac{\partial z}{\partial x} + E \frac{\partial z}{\partial y} + Gz = 0,$$

can always be transformed to (46.7) . A study of (45.4) will take care of (46.8) while a study of (45.9) will take care of (46.4), unless $\bar{E} = 0$ whence (46.4) is essentially an ordinary differential equation.

In the case of (46.1) which reduced to (46.2), if $\bar{e}(\zeta,\eta) \neq 0$, we can perform a similar further simplification and get, instead of (46.7),

$$\frac{\partial^2 u}{\partial s^2} - \frac{\partial u}{\partial \eta} + M(s,\eta) \cdot u + N(s,\eta) = 0.$$

So equations of the form:

$$(46.9) \quad \frac{\partial^2 z}{\partial x^2} - \frac{\partial z}{\partial y} + g(x,y)z + h(x,y) = 0$$

will effectively represent linear equations (46.1).

47. A special case of (46.1) was considered by Feller[1]:

Let a, a_t, a_x, a_{xx} exist and be continuous in R and a > 0.

Let $\varphi(t,x) = \int_0^x \frac{dy}{\sqrt{a(t,y)}}$ and assume $\lim_{x\to-\infty} \varphi(t,x) = -\infty$,

$\lim_{x\to-\infty} \varphi(t,x) = +\infty$. (R: $-\infty<x<+\infty$; $t_0<t<t_1$)

Consider the equation

$$(47.1) \quad M(z) = 0$$

where $M(u) \equiv u_t + \sqrt{a}\,(\sqrt{a}\,u_x)_x - \sqrt{a}\,\varphi_t\,u_x$

and its adjoint equation

$$(47.2) \quad M^*(w) = 0$$

where $M^*(u) \equiv -u_t + (\sqrt{a}\,(\sqrt{a}\,u)_x)_x + (\sqrt{a}\,\varphi_t\,u)_x$.

Then if $t_0<t<\tau<t_1$,

$$(47.3) \quad U_0(t,x;\tau,\zeta) = \frac{1}{2\sqrt{\pi}}\,\frac{1}{\sqrt{a(\tau,\zeta)}}\,\frac{1}{\sqrt{\tau-t}}\,e^{\frac{-\{\varphi(\tau,\zeta)-\varphi(t,x)\}^2}{4(\tau-t)}}$$

is, as a function of t and x, a solution of (47.1), and as a function of τ and ζ a solution of (47.2). Also

$$\lim_{t\to\tau^-} U_0(t,x;\tau,\zeta) = 0 \qquad (x \neq \zeta)$$

U_0 is continuous for $t < \tau$, and all x and ζ. We call it a *fundamental solution* of (47.1).

Consider the equation

$$(47.4) \quad L(z) = 0 \text{ where}$$

$$L(u) \equiv u_t + a(t,x)u_{xx} + b(t,x)u_x + c(t,x)u$$

where a, b, c are defined in R and $a \neq 0$ (so assume a > 0)

Then $L(u) \equiv M(u) + \lambda\,u_x + cu$ where $\lambda = b - \frac{a_x}{2} + \sqrt{a}\,\varphi_t$

Assume also:

(a_1) $a, a_t, a_x, a_{xx}, b, b_x, c$ all exist in R and in the neighborhood of every point (t_0,x_0) satisfy a Lipschitz condition of the form:

$|g(t,x)-g(t_0,x_0)| < K\,(|t-t_0|^\alpha + |x-x_0|^\alpha) \quad (\alpha>0)$

(a_2) a, 1/a, λ, λ_x, c are bounded in R.

Then setting $U_{n+1}(t,x;\tau,\zeta) = \int_t^\tau \int_{-\infty}^{+\infty}\{\lambda(p,q)\dfrac{\partial U_n(p,q;\tau,\zeta)}{\partial q}$ $+ c(p,q)\,U_n(p,q;\tau,\zeta)\}\,U_0(t,x;p,q)\,dq\,dp$ for $n = 0,1,2,\ldots$, where U_0 is defined in (47.3), it is shown that $\sum_{n=1}^{\infty} U_n(t,x;\tau,\zeta) = V(t,x;\tau,\zeta)$ converges absolutely and uniformly. We define

$$U(t,x;\tau,\zeta) = U_0(t,x;\tau,\zeta) + V(t,x;\tau,\zeta).$$

Then $L(U) = 0$, so as a function of t and x, $z = U(t,x;\tau,\zeta)$ is a solution of (47.4). It is called the *fundamental solution* of (47.1).

For every $\delta > 0$, $\lim\limits_{t\to\tau} \int_{-\infty}^{+\infty} U(t,x;\tau,\zeta)\,dx = 1;$

$$\lim_{t\to\tau} \int_{|x-\zeta|>\delta} U(t,x;\tau,\zeta)\,dx = 0$$

The solution of an initial value problem in terms of this fundamental solution is given by:

THEOREM 47.1: Let a, b, c be defined in $R_0 \begin{cases}-\infty<x<+\infty \\ t<T\end{cases}$ and let g(x) be continuous. Let (a_1) and (a_2) hold. Then there is a unique solution $z = \varphi(t,x)$ of $L(z) = 0$ which is bounded in R_0 and such that $\lim\limits_{t\to T^-} \varphi(t,x) = g(x)$. It is given by:

$$\varphi(t,x) = \int_{-\infty}^{+\infty} g(\zeta)\cdot U(t,x;T,\zeta)\,d\zeta$$

where U is the fundamental solution defined above.

[In proof, use the fact that if $L^*(u) = -u_t + (au_{xx}) - (bu)_x + cu$

$= M^*(u) - (\lambda u)_x + cu$, then a fundamental solution is $U^*(t,x;\tau^*,\zeta^*) \equiv U(\tau^*,\zeta^*;t,x)$. Also

$$\int_{t'}^{t''} dt \int_{-\infty}^{+\infty} \{vL(u) - uL^*(v)\}dx = \int_{-\infty}^{+\infty} \{u(t'',x)v(t'',x) - u(t',x)v(t',x)\}dx.]$$

COROLLARY: If $f(t,y)$ is continuous and bounded in R_0, and the rest of the hypotheses of theorem hold, there is a unique bounded solution of:

$$(47.5) \quad L(z) + f(t,x) = 0$$

such that $\lim_{t \to T^-} \varphi(t,x) = g(x)$. It is given by:

$$u(t,x) = \int_t^T dp \int_{-\infty}^{+\infty} f(p,q)\, U(t,x;p,q)\, dq$$

Gevrey[1] considered a reduced form of the linear equation (*) with boundary conditions on a segment of $y = y_0$ together with adjoining curves lying in the halfplane $y > y_0$. The equation is

$$\frac{\partial^2 z}{\partial x^2} - \frac{\partial z}{\partial y} = a\frac{\partial z}{\partial x} + c\frac{\partial z}{\partial y} + f.$$

Using a Green's function, he obtained a representation such as that of the previous theorem. He also established existence of a solution, but there is no brief statement of his theorem which can be given here. (See Gevrey[1], pp. 374-387.)

For other theorems applicable to equation (46.1) see the discussion for n independent variables, and also the theorems of the next section.

48. Suppose (43.3) is of the form:

$$\frac{\partial^2 z}{\partial x^2} + b\frac{\partial z}{\partial y} = \bar{f}(x,y,z,\frac{\partial z}{\partial x})$$

where b is of constant sign; if b < 0 this can be reduced to

$$(48.1) \quad \frac{\partial^2 z}{\partial x^2} - \frac{\partial z}{\partial y} = f(x,y,z,\frac{\partial z}{\partial x})$$

Gevrey[1], pp. 390-394, considered solutions of this equation, obtained as limits of solutions of equations of the kind considered in previous sections. He proved the following for problem D:

THEOREM 48.1: Given (48.1), where $f(x,y,z,p)$, for (x,y,z,p) in R_0: $[(x,y)\ \varepsilon R,\ |z| < A,\ |p| < B]$ satisfies a Lipschitz condition:

$$|f(x,y,z,p) - f(x,y,z_1,p_1)| \leq L|z-z_1| + N\,|p-p_1|$$

and is continuous; given a continuous closed contour C in the interior of R, and such that the equations defining: $x = x(\mu)$, $y = y(\mu)$ satisfy conditions of the form:

$$|x(\mu) - x(\bar{\mu})| < K|\mu - \bar{M}|^{\frac{\alpha+1}{2}} \qquad (0<\alpha\leq 1);$$

given a function $h(x,y)$ defined on C - that is, $h[x(\mu),y(\mu)] = h(\mu)$ which is continuous and has h_x continuous on C. Then there exists a solution of (48.1) in R, the interior of C, which assumed values of $h(x,y)$ on C. [Proof by successive approximation.]

This same equation has been considered by Siddiqi and by Lewis[1]; for a problem of type H, Lewis' result is:

THEOREM 48.2: Given (48.1) where f is defined for (x,y,z,p) in

$$R_0:\ 0\leq x\leq \pi;\ 0\leq y\leq Y;\ |z| \leq Z,\ |p| \leq P.$$

Let f be of class A^1 with respect to p, z, x. Let $g(x)$ be defined for $0 \leq x \leq \pi$, $g(0) = g(\pi) = 0$, $g'(x)$ absolutely continuous; $[g''(x)]^2$ summable. Then there exists a solution $z = \varphi(x,y)$ of (48.1) defined for $0 \leq x \leq \pi$; $0 \leq y \leq Y_0$ such that

$$\varphi(0,y) \equiv \varphi(\pi,y) = 0 \text{ and } \varphi(x,0) = g(x).$$

(It is given in the form: $\varphi(x,y) = \Sigma\, a_k(x) \sin ky$)

Minakshi-Sundaram[2] studied a slightly more general equation than (48.1):

$$(48.2) \quad \frac{\partial u}{\partial t} = \frac{\partial}{\partial x}\left(p \frac{\partial u}{\partial x}\right) - f\left(x, t, u, \frac{\partial u}{\partial x}\right)$$

by exactly the same methods. In a later paper, Sundaram[3] found a solution to a periodic problem with the same boundary conditions using the results of Schauder. Karimov[1] considered periodic solutions for an equation of the same type.

The only existence theorem for the equation (43.3) I have seen is given in Gevrey[1] (pp. 394-402). The conditions are quite involved and will not be given here.

49. An important case of the parabolic equation in n variables is:

$$(49.1) \quad \frac{\partial u}{\partial y} = A_0 u + \sum_{i=1}^{n} A_i \frac{\partial u}{\partial x_i} + \sum_{i,k=1}^{n} B_{ik} \frac{\partial^2 u}{\partial x_i \partial x_k}$$

where $\Sigma\, B_{ik}\, u_i\, u_k$ is a positive definite quadratic form. S. Bernstein[1] considered solutions of (49.1) for problems of the type H_2. That is, he established the existence of a solution $\varphi(y, x_1, \ldots, x_n)$ of (49.1) such that

$$(49.2) \quad \varphi(0, x_1, \ldots, x_n) = g(x_1, \ldots, x_n)$$

and such that φ is bounded for all x_i in $-\infty < x_i < +\infty$. Here g is of class A^3. The method, that of *modules*, is rather involved. Another method of attack is based upon the notion of a "fundamental solution" (see section 46 and the following theorem for an exact definition). For the equation:

$$(49.3)\quad \frac{\partial u}{\partial y} = \Sigma \frac{\partial^2 u}{\partial x_i^2} \equiv \Delta(u)$$

the fundamental solution is

$$(49.3a)\quad u = [4\pi(y-\eta)^{n/2}]\exp[-\frac{\Sigma(x_i-\xi_i)^2}{4(y-\eta)}],\ (y>\eta)$$

Using this as a first approximation, Gevrey[1] established, by successive approximations, the existence of a fundamental solution of:

$$(49.4)\quad \frac{\partial u}{\partial y} = H(u) + \Sigma A_i \frac{\partial u}{\partial x_i} + A_0 u$$

when $H(u) = \Delta(u)$. The self-adjoint equation, which is (49.4) when

$$H(u) = \Sigma \frac{\partial}{\partial x_i}(a_{ik} \frac{\partial u}{\partial x_k})$$

and $a_{ik} = a_{ki}$, can be reduced to the case of Gevrey for $n\geq 2$; for $n=3$, E. Rothe[1] found the fundamental solution of $\frac{\partial u}{\partial y} = H(u)$ by using Green's functions. When $a_{ik} = a_{ik}(x)$ F. Dressel[1], generalized Rothe's results, using methods that stemmed from Levi, to get the fundamental solution for (49.1):

THEOREM 49.1: Let R be an open bounded region in $x = (x_1, x_2, \ldots, x_n)$-space, with boundary R'; let A_0, A_i, $B_{ij,y}$, $B_{ij,xx}$ satisfy Lipschitz conditions of order γ: $|A(x,y) - A(s,t)| \leq N[|y-t|^\gamma + \Sigma |x_i-s_i|^\gamma]$ for $\eta < y-\delta<t<y+\delta$; $x-b\leq s\leq x+b$ where $0<\gamma\leq 1$. Let $B_{ik} = B_{ki}$ and $\Sigma B_{ik}\xi_i\xi_k$ be a positive definite form.

Then there exists a function $\Gamma(x,y;\xi,\eta)$ such that for each x and ξ in R, and for $y > \eta$, Γ is a regular solution of (49.1), and for each g(x) continuous in R+R′ and for $T \subset R$, $\lim_{y\to\eta} \int_T \varphi(\xi)\Gamma(x,y;\xi,\eta)d\xi = g(x)$ when x is in the interior of T but is 0 if x is in the interior of R-T.

This result was given later by Dressel[2] for unbounded regions.

The purpose of fundamental solutions is to get solutions of problems of type H, such as that given on page 175, in terms of an integral which involves Γ and the initial or boundary conditions. See, for example, Rothe's paper, where he used the fundamental solution to solve several problems, including one that involved the non-homogeneous equation $\frac{\partial u}{\partial y} = H(u) + f(x,y)$. For the case of equation (49.3), solution of problems using (49.3a) are classic; recently Heins[1] showed how, in the case of a condition like (49.2), this could be obtained using an iterated Fourier transform.

It is also possible to reduce equation (49.1), by means of a Laplace transform, to an elliptic differential equation in $(x_1, \ldots, x_n)$; see, for example, Doetsch[1].

A non-linear equation,

$$(49.5)\quad \frac{\partial z}{\partial y} = K\left(\frac{\partial^2 z}{\partial x_1^2} + \frac{\partial^2 z}{\partial x_2^2}\right) + f(x_1, x_2, y, z)$$

was considered by Sundaram[3], with special conditions on f.

Since any linear equation in n variables which has second derivatives involving fewer than n variables is called parabolic, there are other kinds than (49.1).

Piscounov[1] considered *ellipticc-parabolic* equations of the type:

$$(49.6)\quad \frac{\partial^2 z}{\partial x^2} = \sum_{i=1}^{n} a_i \frac{\partial z}{\partial y_i} + c\frac{\partial z}{\partial x} + bz + f$$

where a_i, c, b, f are functions of x, $y_1, \ldots, y_n$ and $\sum_{i=1}^{n} a_i^2 \geq \delta > 0$.

First consider a domain V in the (n+1) dimensional space of $x, y_1, \ldots, y_n$ that is bounded by the surfaces:

$$\phi_i(y_1, \ldots, y_n) = 0 \qquad (i=1,2)$$

$$x = \varphi_j(y_1, \ldots, y_n) \qquad (j=1,2)$$

where all four surfaces are bounded, where ϕ_1, φ_1, φ_2 are of class A^3, where $\varphi_1-\varphi_2 \geq \delta$ everywhere in the domain bounded by $\phi_1 = 0$, and $\phi_2 = 0$, though it may vanish on $\phi_2 = 0$, and where lines β - (that is, families of integral curves of the system of ordinary differential equations:

$$y_i' = a_i(x, y_1, \ldots, y_n) \quad (i=1,\ldots,n))$$

which are in V are not tangent to $\phi_1 = 0$ and do not cut the surface $\phi_2 = 0$. Then the following results hold:

THEOREM 49.2: Let a_i, b, c, f be functions of class A^3 in V; let $f^{\nu}(x, y_1, \ldots, y_n)$ $(\nu=1,2,3)$ be defined on the surfaces $\phi_1 = 0$, $x = \varphi_2$, $x = \varphi_1$ respectively and let all three be of class A^3 (except that f' may be continuous only with respect to x and of class A^3 with respect to $y_1, \ldots, y_n$). Then there exists a unique function $\varphi(x, y_1, \ldots, y_n)$ such that $z = \varphi(x, y_1, \ldots, y_n)$ is a solution of (49.6) in the domain V which coincides with f_1, f_2, f_3 on the surfaces $\phi_1 = 0$, $x = \varphi_2$ and $x = \varphi_1$ respectively.

THEOREM 49.3: Let a_i, b, c, f be functions of class A^3 in V; let $U_0(x, y_2, \ldots, y_n)$, $U_1(y_1, \ldots, y_n)$ $U_2(y_1, \ldots, y_n)$ be defined on $\phi_1 = 0$, $x = \varphi_1$, $x = \varphi_2$ respectively. Then there exists a unique function $\phi(x, y_1, \ldots, y_n)$ which is of class A^1 in $y_1, \ldots, y_n$ and of class A^2 in x, such that $z = \phi(x, y\ , \ldots, y_n)$ is a solution of (49.6) and such that

$$\phi(x, 0, y_2, \ldots, y_n) \equiv U_0(x, y_2, \ldots, y_n),$$

$$\phi_x(\varphi_1, y_1, \ldots, y_n) \equiv f_1(y_1, \ldots, y_n),$$

$$\phi_y(\varphi_1, y_1, \ldots, y_n) \equiv f_2(y_1, \ldots, y_n).$$

Assume that V is bounded by a continuous S_1 of (n-1) dimensions containing no segments parallel to the x axis.

H. Elliptic Equations

50. For the equation

$$(50.1) \quad a(x,y)\frac{\partial^2 z}{\partial x^2} + 2b(x,y)\frac{\partial^2 z}{\partial y \partial x} + c(x,y)\frac{\partial^2 z}{\partial y^2} + f(x,y,z,\frac{\partial z}{\partial x},\frac{\partial z}{\partial y}) = 0 \qquad (ac>b^2)$$

where a, b, c are continuous in R_0, $ac>b^2\geq 0$ implies $a\neq 0$ and $c\neq 0$ in R_0, so we can take $a>0$ and $c>0$ with no loss of generality. We saw on page 103 that a transformation will reduce (50.1) to a simpler form: (dividing by $A(\xi,\eta)$, which is not 0).

$$\frac{\partial^2 \omega}{\partial \xi^2} + M\frac{\partial^2 \omega}{\partial \eta^2} + \overline{F}(\xi,\eta,\omega,\omega_\xi,\omega_\eta) = 0.$$

Since $M>0$, by another transformation this will become:

$$(50.2) \quad \frac{\partial^2 \omega}{\partial \xi^2} + \frac{\partial^2 \omega}{\partial \eta^2} + \overline{F}(\xi,\eta,\omega,\omega_\xi,\omega_\eta) = 0.$$

Since there are no real characteristics, a problem like problem G for hyperbolic equations has no meaning for (50.1). On the other hand problem C can be formulated for any curve C_0. Indeed, from the theorems in section I, if a,b,c,f are functions of class A^∞, and the same is true of the initial data, there always exists a solution of class A^∞ of problem C in the neighborhood of each point of the curve. A natural generalization of this theorem would be to find a solution of problem C which is not analytic, or when f or the initial data is not analytic. But it turns out that if a,b,c,f are analytic, then every solution of (50.1) is analytic. This theorem, called Hilbert's theorem, was proved by H. Lewy for (50.2) in Gott. Nachr.(1929). It may be more accessible in the appendix to Hadamard[2].

The accurate statement for the most general elliptic equation is given below in section F. For the proof in the simplest case of equation (51.1) below, see Tamarkin and Feller[1], p. 18.

Thus the problems usually studied for elliptic equations are quite different from those considered for hyperbolic equations. Three basic problems are the following:

Problem D: (Dirichlet) Let G be a bounded region and Γ its boundary. Let f(s) be continuous on Γ. It is required to establish the existence of a function $\varphi(x,y)$ of class A^2 in $G+\Gamma$ such that $z = \varphi(x,y)$ satisfies (50.1) in G and such that $\varphi(x,y)$ coincides with f(s) on Γ.

When f(s) = 0, we shall call it problem D_0.

A variation of problem D is to require that $\varphi(x,y)$ be of class A^2 only in G and that $\lim \varphi(x,y) = f(s)$, as $(x,y)\to\Gamma$. Call this D_1.

Problem N: (Neumann) Let G be a bounded region with boundary Γ and let h(s) be continuous on Γ and $\int_\Gamma h(s)ds = 0$. It is required to establish the existence of a function $\varphi(x,y)$ of class A^2 in $G+\Gamma$ such that $z = \varphi(x,y)$ satisfies (50.1) in G and such that along Γ, $h(s) = \frac{\partial\varphi}{\partial n}$, the directional derivative in the direction of the inner normal.

(Variations of this analogous to those given above will be called N_0 and N_1.)

Problem T: Let G, Γ, f(s), h(s) be defined as in problems D and N. It is required to establish the existence of a function $\varphi(x,y)$ of class A^2 in which satisfies (50.1) in G and such that along Γ,

$$\frac{\partial\varphi}{\partial n} = h(s)\cdot\varphi + f(s).$$

51. In this section we consider the special elliptic equations:

(51.1) $\frac{\partial^2 z}{\partial x^2} + \frac{\partial^2 z}{\partial y^2} = 0$ (51.3) $\frac{\partial^2 z}{\partial x^2} + \frac{\partial^2 z}{\partial y^2} + cz = 0$

(51.2) $\frac{\partial^2 z}{\partial x^2} + \frac{\partial^2 z}{\partial y^2} = f(x,y)$ (51.4) $\frac{\partial^2 z}{\partial x^2} + \frac{\partial^2 z}{\partial y^2} + cz = f(x,y)$

and their generalizations to n dimensions:

$$(51.5)\quad \sum_{k=1}^{n} \frac{\partial^2 z}{\partial x_k^2} = 0 \qquad (51.7)\quad \Sigma \frac{\partial^2 z}{\partial x_k^2} + cz = 0$$

$$(51.6)\quad \Sigma \frac{\partial^2 z}{\partial x_k^2} = f(x_1,\ldots,x_n) \qquad (51.8)\quad \Sigma \frac{\partial^2 z}{\partial x_k^2} + cz = f(x_1,\ldots,x_n)$$

It was shown in section A that any linear elliptic equation with constant coefficients, except for the term not involving z, could be reduced to one of the above forms. These equations arose in connection with the study of potential problems. There is a very large literature in connection with them, for their solutions have many interesting properties, but we are concerned here only with some basic existence theorems.

One classic line of reasoning, which formed the basis of the belief that was current for a long time, that problem D for (51.1) or (51.5) always had a solution was the following: (Results only summarized here; proofs found in any standard work, such as Horn[1], Courant-Hilbert[1], or Kellogg[1]). If R is a region whose boundary consists of a finite number of rectifiable arcs, and u and v are functions of class A^2 in R, Green's formula says that:

$$\iint_R (u\Delta v - v\Delta u)\,dS = \int_c \left(v\frac{\partial u}{\partial n} - u\frac{\partial v}{\partial n}\right) ds$$

If (ξ,η) is a fixed point outside R then setting $r = \sqrt{(x-\xi)^2+(y-\eta)^2}$, $V=-\log r$ is a regular solution - that is, a solution of class A^2 - of (51.1) in R; if (ξ,η) is in R, V is still a solution in R, for all $(x,y) \neq (\xi,\eta)$. If u is a regular solution in R, $\int_c \frac{\partial u}{\partial n}\,ds = 0$. If u and $\frac{\partial u}{\partial n}$ both have prescribed values on the boundary C, and u is to be a regular solution inside R, it is uniquely determined:

$$u(\xi,\eta) = -\frac{1}{2\pi}\int_C \left[V\frac{\partial u}{\partial n} - u\frac{\partial V}{\partial n}\right] ds$$

But usually, as in problem D, only the values of u along C are prescribed. However, suppose one can find $\Omega(x,y;\xi,\eta)$ which assumes the value log r on C and is a regular solution in R; then $G(x,y;\xi,\eta) = -\log r + \Omega$ is a Green's function" - that is, a solution of (51.1) inside R except at the point (ξ,η), which vanishes on C. In that case, the value of u inside R is determined and is given by:

$$(51.9)\quad u(\xi,\eta) = \frac{1}{2\pi}\int_C u\,\frac{\partial G}{\partial n}\,ds = \frac{1}{2\pi}\int_C f(s)\,\frac{\partial G}{\partial n}\,ds$$

This is not an existence theorem; it simply states that if R is a region for which a Green's function can be defined, and u is a solution to problem D in R - that is, a regular solution of (51.1) inside R which becomes f(s) on the boundary C - then inside R, u is given by formula (51.9) and hence is unique. Ω is called a fundamental solution.

Thus the problem is reduced to establishing a Green's function for R. For a circular region there is an explicit formula for G; if C is a circle of radius a about the origin, and H is a point with coordinates (ξ,η), whose inversion in the circle C is $\bar{H}$, let the coordinates of $\bar{H}$ be $(\bar{\xi},\bar{\eta})$ $\quad (OH \cdot O\bar{H} = a^2)$. Then if

$$r^2 = (x-\xi)^2 + (y-\eta)^2,\ h^2 = \xi^2 + \eta^2,\ \bar{\Omega}^2 = (x-\bar{\xi})^2 + (y-\bar{\eta})^2$$

the Green's function for the region R bounded by the circle C is: $G(x,y;\xi,\eta) = \log \frac{h\,\bar{r}}{a\,r}$. Since the polar coordinates of $\bar{H}$ are (h,0) and of P are (r,φ), formula (51.9) becomes:

$$u(\xi,\eta) = \frac{1}{2\pi}\int_0^{2\pi} f(a\varphi)\left\{\frac{a^2-h^2}{a^2-2ah\cos(\varphi-\theta)+r^2}\right\}\,d\varphi\,,$$

the Poisson integral. Thus we do have the existence of a regular solution to problem D for the case of this region. More generally, we have:

THEOREM 51.1: Let G be any region with boundary Γ which can be mapped into the interior R of a circle of radius a, so that Γ goes into C. (The map must be conformal). Then the Green's function exists for such a region and hence by (51.9) there is a regular solution of problem D, for (51.1).

Another classic result, in terms of the Green's function defined above is for the non-homogeneous equation (51.2): There is a unique function which is a regular solution of (51.2) inside R and vanishes on C; it is given by:

$$u(x,y) = \frac{1}{2\pi} \iint_G G(x,y;\xi,\eta)\ f(x,y)dxdy$$

Courant, Friedrichs and Lewy[1] in a long paper describing the solution of many of the simple fundamental equations which occur in applications by means of difference equations, consider problem D for both (51.1) (Equation of Laplace) and (51.2) (Equation of Poisson). Thus they prove, for problem D:

THEOREM 51.2: Given (51.2), where $f(x,y)$ is of class A^2 in a region R, let G be a simply connected region where boundary is formed by a finite number of arcs, each with a continuously turning tangent, and such that G lies in the interior of R. Let $g(x,y)$ be a function defined on the boundary of G. Then there exists a function $\varphi(x,y)$ such that $z = \varphi(x,y)$ is a solution of (51.2) in G and $\varphi(x,y) \equiv g(x,y)$ on the boundary of G. Furthermore, $\varphi(x,y) = \lim_{h\to 0} \varphi_h(x,y)$ where $\varphi_h(x,y)$ is the solution of the corresponding problem for the difference equation $Z_{x\bar{x}} + Z_{y\bar{y}} + f = 0$ on a mesh of width h.

One can also get the solution in terms of Green's functions, by combining two formulas on the preceding page:

$$(51.10) \quad z(x,y) = \int_c f(s) \frac{\partial G}{\partial n}\ ds - \frac{1}{2\pi} \iint_G G(x,y;\xi,\eta)\ f(x,y)\ dx\ dy.$$

Of course, solutions of this nature depend upon a **determination** of $G(x,y;\xi,\eta)$ and that depends upon the shape of the region. One can always construct Green's functions for the regions of the theorem but for practical problems, this may be more complicated than applying the method given in Courant, Friedrichs and Lewy[1].

Problems N and T can also be solved by these methods.

Southwell[1] has done much in the evaluation of solutions of the problems on page 180 both for (51.1) and (51.2). His *relaxation* methods are essentially those of difference equations. They have been extended by Fox[1] and Emmons[1]. Indeed, Akushsky[1] has a method of solving problem D in a rectangular region by using punch card machines.

For $n > 2$, analogous results hold, except that $-\log r = \gamma(r)$ is replaced by $\gamma(r) = \frac{1}{\omega_n(n-2)} r^{2-n}$. But again the Green's function for G is the solution of (51.5) which is regular at all points P: $(x_1, \ldots, x_n)$ in G except P = Q: $(\xi_1, \ldots, \xi_n)$, is continuous in $G+\Gamma$ and vanishes on Γ. ω, the fundamental solution, is a regular solution in R, which becomes $-\gamma(r)$ on C; then the Green's function is:

$$G(P,Q) = \gamma(r) + \omega$$

It has the property that $G(P,Q) = G(Q,P)$. Then formula (51.9) gives the solution to problem D for (51.5), provided the Green's function exists, and

$$u(P) = \int_C f(s) \frac{\partial G}{\partial n} ds \frac{1}{(2\pi)^n} \int_G G(P,Q) \cdot g(Q)\, dQ$$

is the solution to problem D for (51.6).

Raynor[1] gives an excellent summary of the classical solutions of D when R is a sphere, etc. and of other properties of harmonic functions, for the case $n = 3$. We state one of his results:

THEOREM 51.3: Problem D has a solution for every 3-dimensional region R where boundary B is such that, if a sphere be drawn about any point of B, the ratio of measure of the points of the surface of the sphere on B to the whole area of the sphere will remain < 1 as radius of sphere $\to 0$. (Shrinking process need not be continuous, but may be made in a denumerable number of steps.)

The problem of Dirichlet is not possible for all regions - it is now known that there does not always exist a function V taking on given values at <u>all</u> points of the frontier. In 1924, Wiener defined a frontier point as regular if V is continuous and takes on the value of f at that point; otherwise it is irregular, and gave a way of distinguishing between them. A domain for which the problem of Dirichlet is possible he called a normal domain. He then generalized the problem and established the following result:

THEOREM 51.4: Let r be an open set (bounded or not), Σ its frontier (bounded) and f(P) a continuous function on Σ. Let F be a continuous function in the entire space and which coincides with f on Σ, and r_n a sequence of normal domains interior to r and converging to r. If V_k is a solution of problem D in r_k for values of F on its frontier Σ_k, then the sequence of functions V_k converges uniformly to a function V, which is harmonic, in every closed subregion of V. This function is independent of the choice of r_k and of F.

In his talk before the 1935 Congress, Vasilesco[1] summarized other classical methods. The Neumann and Dirichlet problems for (51.4) were given by Giraud[1] who extended this to n dimensions.

Problem D for (51.3) has been solved by finite differences, using a three-dimensional net by Mikeladze[1]. Southwell[1] also considers such equations. A recent study of problem D for $\Delta z = 0$ where $\Delta u = \frac{\partial^2 u}{\partial x_1} + \cdots + \frac{\partial^2 u}{\partial x_n}$ by finite differences was given by Petrowsky[1]. According to the reviews it is very simple. (Paper is in Russian.)

52. The linear differential equation of elliptic type can be represented, we saw in Section 50, by

$$(52.1)\quad \frac{\partial^2 z}{\partial x^2} + \frac{\partial^2 z}{\partial y^2} + d\frac{\partial z}{\partial x} + e\frac{\partial z}{\partial y} + gz = h(x,y).$$

If we set $L(u) \equiv u_{xx} + u_{xy} + du_x + eu_y + gu$, this equation is

$$L(u) = h(x,y).$$

The homogeneous equation is:

$$L(u) = 0.$$

One way of establishing existence of a solution is to make it depend upon the existence of a Green's function $K(x,y;\xi,\eta)$ of (50.3) $\Delta u = 0$ where $\Delta u \equiv \frac{\partial^2 u}{\partial x^2} + \frac{\partial^2 u}{\partial y^2}$. Thus we have for problem D:

THEOREM 52.1: Let G be a region such that $K(x,y;\xi,\eta)$ exists - for example, let G be a region which can be mapped onto a circle, and let

$$|K_x(x,y;\xi,\eta)| < \frac{c}{r}; \quad |K_y(x,y;\xi,\eta)| < \frac{c}{r}$$

where $r = \sqrt{(x-\xi)^2+(y-\eta)^2}$ and c is independent of x,y,ξ,η. Let d, e, g and h be of class A^1 in G. Then the equation

$$L(z) = h(x,y)$$

has a solution $z = \varphi(x,y)$ which vanishes on the boundary Γ of G, provided G is sufficiently small.

Proof: (By integral equations) We are looking for a solution of type:

$$z = \int_G\int K(x,y;\xi,\eta)\ \rho(\xi,\eta)\ d\xi d\eta$$

where ρ is of class A^1 in G. By hypothesis $H(x,y;\xi,\eta) = aK_x+bK_y+cK$ satisfies:

$$|H(x,y;\xi,\eta)| > \frac{\alpha}{r}.$$

Now $L(z) = \int\int H\rho(\xi,\eta)\ d\xi d\eta - \rho(x,y)$ so we have reduced problem to solving integral equation

$$\rho = -f + \int_G\int H(x,y;\xi,\eta)\ \rho(\xi,\eta)\ d\xi d\eta$$

for ρ. By Fredholm theorem, there is a solution of an equivalent integral equation, for small enough G - so small that

$$\int_G\int\int_G\int H^2(x,y;\xi,\eta)\ dxdyd\xi d\eta < 1.$$

(This proof can be found in several places. See Courant-Hilbert[1], vol. 2, for example.)

One can find *fundamental solutions* or *elementary solutions* in the same way. Such a solution $\gamma(x,y;x_0,y_0)$ would be one which is a solution of (52.1) everywhere except at (x_0,y_0). It is of the form:

$$\gamma(x,y;x_0,y_0) = \log r_0 + \iint_G (-\log r)\,\rho(\xi,\eta)\,d\xi\,d\eta$$

where $r = \sqrt{(x-\xi)^2+(y-\eta)^2}$ and (x_0,y_0) is a fixed point inside G, with $r_0 = \sqrt{(x-x_0)^2+(y-y_0)^2}$. One gets $\rho(x,y)$ by solving the integral equation:

$$\rho(x,y) = \iint_G K(x,y;\xi,\eta)\,\rho(\xi,\eta)\,d\xi\,d\eta + \frac{1}{2\pi}(L(-\log r)-f).$$

(See Courant-Hilbert)

However, there are many other ways of treating (52.1). One can use the method of calculus of variations, where the problem of finding a solution of an equation which $\equiv 0$ on the boundary (problem D_0) is reduced to a problem of making a certain function a minimum. Schauder has established existence by functional operators. (See section 53.) Tautz[1] has also used general methods on 52.1. Most recently, we have direct numerical methods applied to the solution of problem D and problem D_0. Vecoua[1] and Mikeladze[2], have used finite differences to solve problem D for (52.1), and give actual upper bound of errors in approximating solutions. Chernoff[1], following Bergmann[1], considered z as the real fact of a certain complex function.

53. Non-linear equations of elliptic type

$$(53.1)\quad F\left(x,y,z,\frac{\partial z}{\partial x},\frac{\partial z}{\partial y},\frac{\partial^2 z}{\partial x^2},\frac{\partial^2 z}{\partial x\partial y},\frac{\partial^2 z}{\partial y^2}\right) = 0 \qquad (F_rF_t - 4F_s^2 > 0)$$

were considered in 1910 by S. Bernstein[1]. For the case

$$(53.2) \quad L_1(u) \equiv A(x,y,z,\frac{\partial z}{\partial x},\frac{\partial z}{\partial y})\frac{\partial^2 z}{\partial x^2} + 2B(x,y,\ldots)\frac{\partial^2 z}{\partial x\partial y} + C(x,y,\ldots)\frac{\partial^2 z}{\partial y^2} = 0$$

where A,B,C are analytic functions of x,y,z,p,q, he showed that a solution is always possible to problem D: finding a function which satisfies (53.2) inside a region R and assumes prescribed values on the closed contour B which forms the boundary of R, (provided B is of a certain type). He also showed that if F is analytic, a solution to problem D for (53.1) is possible if one can introduce a parameter α such that for $\alpha = 0$, $F(x,y,z,p,q,r,s,t;\alpha)$ becomes a function for which a solution to problem D is known to exist, while for $\alpha = 1$, it becomes the given function, and such that for $0 < \alpha < 1$ a solution and all its derivatives up to a certain order are bounded. In a later paper (Bernstein[2]) these results and similar ones are restated and clarified. (Assume $F_rF_z < 0$.) Others made his hypotheses less strong; as an example of the type of result, we give the following theorem, due to Deuffer[1]:

THEOREM 53.1: Let Γ be a simply connected region of the x-y plane with an analytic boundary curve. Let $z_0(x,y)$ be defined in $\bar{\Gamma}$ and be of class A^4 there. Let $F(x,y,z,p,q,r,s,t;\alpha)$ be defined for all (x,y) in Γ and to every $z_0(x,y)$ for

$$\left|r - \frac{\partial^2 z_0}{\partial x^2}\right| \le \varepsilon, \quad \left|s - \frac{\partial^2 z_0}{\partial x\partial y}\right| \le \varepsilon, \ldots, \quad |z-z_0| \le \varepsilon, \quad \alpha \le \varepsilon.$$

In the nine-dimensional region B let F, F_x, F_y be bounded and of class A^∞. Let $4F_rF_t - F_s^2 \ge k > 0$ and $F_zF_r \le 0$ in B. Let $F(x,y,z,p,q,r,s,t;0) = 0$ when $z = z_0$, $p = z_{0,x}, \ldots, t = z_{0,yy}$. Then there is an ε^* and a function $z(x,y;\alpha)$ of class A^2 with respect to x and y which for every $|\alpha| \le \varepsilon^*$ is a solution of

$$F(x,y,z,\frac{\partial z}{\partial x},\frac{\partial z}{\partial y},\frac{\partial^2 z}{\partial x^2},\frac{\partial^2 z}{\partial x\partial y},\frac{\partial^2 z}{\partial y^2};\alpha) = 0$$

and which coincides with $z_0(x,y)$ on the boundary of Γ.

Giraud[1] also considered the non-linear equation; we give an example of his results (they are given for n independent variables) for problem D:

THEOREM 53.2: Let F be a function of class A^{∞} for $(x,y)\varepsilon$ R and $(z,p,\ldots,t)\varepsilon S$; let R and S contain (0,0) and (0,...,0) respectively and let $F_z < 0$ in $R \times S$. Let $z \equiv 0$ be a solution of (53.1). Then if on a regular analytic contour C, interior to a circle of radius d in R, an analytic sequence of values: $s\varphi(\alpha)$ is given, where α fixes the position of a point on C and s is a coefficient variable, there exists a unique solution of (53.1) for s sufficiently small, which is of class A^{∞} inside and on C and assumes the values $s\varphi(\alpha)$ on C.

Proof: Expand F is powers of $z,p,\ldots,t$, and neglecting terms of degree higher than one, solve the resulting linear equation, with the given boundary conditions; by induction, having z_n, replace z by z_n+h in (53.1), neglect terms of order >1 in h and the derivatives, find solution of resulting linear equation in h which is 0 on C; set $z_{n+1} = z_n+h$. (This is an extension of the method due to Picard[1], who was one of the first to consider equations of this kind.) Then $\{z_n\}$ tends to a limit, which is shown to be the required solution.

Schauder[3] showed that a sufficient condition for a solution of problem D for (53.2), where the coefficients are of class A is that the non-homogeneous equation $\Delta_1(u) = g(x,y)$ have at most one solution taking on the prescribed value on the boundary, for arbitrary g; the proof is based on his fixed point theorem. By the same means he showed[4] that if A,B,C,G are bounded and continuous for all x,y,z,p,q there exists a solution of

$$(53.3)\quad A\frac{\partial^2 z}{\partial x^2} + 2B\frac{\partial^2 z}{\partial x\partial y} + C\frac{\partial^2 z}{\partial y^2} = G(x,y,z,\frac{\partial z}{\partial x},\frac{\partial z}{\partial y})$$

which is continuous, and has first and second derivatives almost everywhere (he calls this a generalization of problem D). Leray[1] in an excellent discussion of the non-linear problem before the 1935 Conference, generalized Schauder's fixed point theorem, and used it to show how a solution of problem D for (53.3) reduces to finding a majorant for z, and its first and second derivatives, but he does not give any existence theorems.

Simonoff[1] has established the following result:

THEOREM 53.3: Given

$$(53.1)\quad F(x,y,z,\frac{\partial z}{\partial x},\frac{\partial z}{\partial y},\frac{\partial^2 z}{\partial x^2},\frac{\partial^2 z}{\partial x\partial y},\frac{\partial^2 z}{\partial y^2}) = 0$$

where $F(x,y,z,p,q,r,s,t)$ is of class A^1 for $|x| < B$, $|y| < B$ and all p,q,r,s,t, and where there are constants $k_1 > 0$, $k_2 > 0$, $k_3 > 0$ such $4F_rF_t - F_s^2 \geq k_1$, $F_r > 0$; $|F_z| \leq k_2,\ldots,F_t \leq k_2$; $F_z \leq -k_3$. Then in any normal region there is a unique solution to problem D.

Lewy[3] has shown that if F is a regular analytic function in a complex neighborhood N of the real point $(x_0,y_0,\ldots,t_0)$ and if C $x=x(\mu)$, $y=y(\mu)$, $y_0)$ a curve in the (x,y) plane defined in the neighborhood of N_1 of (x_0,y_0) and $z_0(\mu),\ldots,t_0(\mu)$ are functions defined in the neighborhood of μ_0 and satisfying strip equations, then solution to the Cauchy problem is necessarily analytic; when C is not inside N, but part of the boundary, Lewy[3] showed the solution was unique. Hoff established a similar result.

Carleman, Rado, and others considered the important question of the analytic nature, or functional character of solutions, as well as their uniqueness, but such results are not existence theorems, in the strict sense.

54. The n-dimensional equation corresponding to (53.1) is:

$$(54.1)\quad \sum_{i,k=1}^{n} a_{ik} \frac{\partial^2 z}{\partial x_i \partial x_k} + \sum_{j=1}^{n} b_j \frac{\partial z}{\partial x_j} + c\, z \equiv d$$

where $\Sigma\, a_{ik}\, z_i z_k$ is a positive definite quadratic form. One has problems exactly like D and N for this equation. Giraud[1,2,3] has studied this intensively; the general method is to get the solution in terms of solutions of

$$(54.2)\quad \Sigma\, a_{ij} \frac{\partial^2 z}{\partial x_i \partial x_k} = g^2 u.$$

This, in turn, for boundary conditions which are linear combinations of z and $\frac{\partial z}{\partial x_j}$ on Γ, the boundary of a closed region G, is reduced to solution of integral equations of the Fredholm type, such as described in section 52. Gevrey[1] has also done this.

Uniqueness of solution of problem D for (54.1) is easy to show, at least for the homogeneous equation, if one wants a continuous solution, since any solution of the homogeneous equation which vanishes on the boundary vanishes inside.

Schauder[5] has established existence of solution of problem D for (54.1) when the coefficients and the data satisfy certain *Holder* conditions. He reduces (54.1) to

$$(54.3)\quad \Sigma\, a_{ij} \frac{\partial^2 z}{\partial x_i \partial x_j} = f$$

and then considers a similar equation with a_{ij} replaced by a^{λ}_{ij}, a set of functions which reduces to (54.3) for $\lambda = 0$ and to Poisson's equation (52.6) for $\lambda = 0$. Then as with other problems, he obtains bounds that enable him to use his fixed point theorem.

Trjitzinsky[1] has also studied equation (54.1) getting conditions for solution of problem D, when the coefficients are singular.

I. General Second Order Partial Differential Equations

55. When a second order differential equation:

$$(22.2)\quad F\left(x,y,z,\frac{\partial z}{\partial x},\frac{\partial z}{\partial y},\frac{\partial^2 z}{\partial x^2},\frac{\partial^2 z}{\partial x\partial y},\frac{\partial^2 z}{\partial y^2}\right) = 0$$

can be reduced to normal form, (23.8), or (23.8), and the data is analytic, the general Cauchy-Kowalewski theorem gives existence theorems for two specific problems.

THEOREM 55.1: (a) Let $g(y)$ and $h(y)$ be two functions of class A^∞ in a region: $|y-y_0| < \delta$. Set $g(y_0) = z_0$, $g'(y_0) = q_0$, $g''(y_0) = t_0$, $h(y) = p_0$, $h'(y) = s_0$. Let x_0 be arbitrary.

(b) Let $F(x,y,z,p,q,r,s,t)$ be of class A^∞ in a neighborhood N of P: $(x_0,y_0,z_0,p_0,q_0,r_0,s_0,t_0)$ and let $F = 0$ and $F_r \neq 0$ at P.

Then there exists one and only one function $\varphi(x,y)$ which has the properties:

(α) $\varphi(x,y)$ is of class A^∞ in a region R: $|x-x_0| < \delta_1$, $|y-y_0| < \delta_2$.

(β) $z = \varphi(x,y)$ is a solution of the equation (22.2) in R.

(γ) For $|y-y_0| < \delta_2, \varphi(x_0,y) \equiv g(y)$ and $\varphi_x(x_0,y) \equiv h(y)$.

Proof: In a neighborhood of P^0, (22.2) is equivalent to (23.8), which is equivalent to the system in two unknowns:

$$\frac{\partial z}{\partial x} = z_1;\ \frac{\partial z}{\partial x} = f\left(x,y,z,z_1,\frac{\partial z}{\partial y},\frac{\partial z_1}{\partial y}\right)$$

to which the theorems of chapter I can be applied. A direct proof can be found in Horn[1], pp. 192-195.

In certain instances, this can be used to give a result to the problem of Cauchy:

THEOREM 55.2: (a) Let $C_0: x = x_0(\mu)$, $y = y_0(\mu)$ be a given curve and let $z_0(\mu), p_0(\mu), q_0(\mu)$ be given where all 5 functions are of class A^∞ and satisfy:

$$z_0'(\mu) = p_0(\mu) \cdot x_0'(\mu) + q_0(\mu) \cdot y_0'(\mu)$$

for $\mu \varepsilon M$: $|\mu - \mu_0| < \delta$. Set $x^0 = x_0(\mu_0), \ldots, p^0 = p_0(\mu_0)$ and let r_0, s_0, t_0 to be numbers such that

$$F(x^0, y^0, z^0, p^0, q^0, q^0, r^0, s^0, t^0) = 0$$

$$r^0 \cdot x'(\mu_0) + s^0 \cdot y_0'(\mu^0) = p_0'(\mu^0)$$

$$s^0 \cdot y_0'(\mu^0) + t^0 \cdot z_0'(\mu^0) = q_0'(\mu^0)$$

(b) Let $F(x,y,z,p,q,r,s,t)$ be of class A^∞ in a neighborhood of $P^0(x^0, y^0, \ldots, t^0)$ and let

$$\Delta = Ry_0'^2 - Sx_0'y_0' + Tx_0'^2 \neq 0$$

at P_0, where, as in section 22, $R = F_r$, $S = F_s$, $T = F_t$.

Then there exists a unique function $\varphi(x,y)$ such that

(α) $\varphi(x,y)$ is of class A^∞ in R_{δ_1}: $|x - x^0| < \delta_1$, $|y - y^0| < \delta_1$.

(β) $z = \varphi(x,y)$ satisfies (22.2) in R.

(γ) For $|\mu - \mu^0| < \delta_2$,

$$\varphi(x_0(\mu), y_0(\mu)) = z_0(\mu); \quad \varphi_x(x_0(\mu), y_0(\mu)) = p_0(\mu);$$

$$\varphi_y(x_0(\mu), y_0(\mu)) = q_0(\mu).$$

Proof: By the implicit function theorem, the equations:

$$F(x_0(\mu), \ldots, p_0(\mu), r, s, t) = 0$$

$$r \cdot x'(\mu) + s\, y_0'(\mu) = p_0'(\mu)$$

$$s \cdot x_0'(\mu) + t\, z_0'(\mu) = q_0'(\mu)$$

will have a simultaneous solution $r = r_0(\mu)$, $s = s_0(\mu)$, $t = t_0(\mu)$ such that $r^0 = r_0(\mu_0)$, $s^0 = s_0(\mu_0)$, $t^0 = t_0(\mu_0)$, since the Jacobian:

$$\begin{vmatrix} R & S & T \\ x_0'(\mu) & y_0'(\mu) & 0 \\ 0 & x_0'(\mu) & y_0'(\mu) \end{vmatrix} = \Delta \neq 0$$

Thus we have defined for a neighborhood of P^0 an integral strip of second order - that is, a set of functions satisfying (24.4) and (24.6).

Since $\Delta \neq 0$, either $x_0' \neq 0$ or $y_0' \neq 0$; suppose $y_0'(\mu) \neq 0$. Then $y = y_0(\mu)$ has an inverse $\mu = \mu_0(y)$ with $\mu_0 = \mu_0(y_0)$; let $d(y) = x_0(\mu_0(y))$; $g(y) = z_0[\mu_0(y)]$; $h(y) = p_0[\mu_0(y)]$. By the strip equation, $q_0[\mu_0(y)] = g'(y) - h(y)\dfrac{x_0'[\mu_0(y)]}{y_0'[\ldots]} = g' - h\cdot d'$. The transformation T: $\bar{x} = x-d(y), \bar{y} = y$ has Jacobian 1 and inverse T′: $x = \bar{x} + d(\bar{y})$, $y = \bar{y}$. Then (see page 22 and pages 99 and 100) if φ were a solution of (22.2), and $\bar{\varphi}(\bar{x},\bar{y}) = \varphi(\bar{x}+d(\bar{y}),\bar{y})$,

$$\varphi_x(\bar{x}+d(\bar{y}),\bar{y}) = \bar{\varphi}_{\bar{x}}(\bar{x},\bar{y}) \quad ; \quad \varphi_{xx} = \bar{\varphi}_{\bar{x}\bar{x}} \quad ; \quad \varphi_{xy} = \bar{\varphi}_{\bar{x}\bar{y}} - d'(\bar{y})\cdot \bar{\varphi}_{\bar{x}\bar{x}}$$

$$\varphi_y = \bar{\varphi}_{\bar{y}} - d'\bar{\varphi}_{\bar{x}} \quad ; \quad \varphi_{yy} = \bar{\varphi}_{\bar{y}\bar{y}} - 2(d'(\bar{y}))\,\bar{\varphi}_{\bar{x}\bar{y}} + [d'(\bar{y})]^2\,\bar{\varphi}_{\bar{x}\bar{x}}$$

So we define: $\bar{x} = x$, $\bar{y} = y$, $\bar{z} = z$, $\bar{p} = p$, $\bar{q} = q+d(y)$, $\bar{r} = r$, $\bar{s} = s+d(y)$, $\bar{t} = t+2d'(y)\,s+d'^2 r + d''\cdot p$ and $\bar{F}(\bar{x},\ldots\ldots,\bar{t}) \equiv F(\bar{x},\bar{y},\bar{z},\bar{p},\bar{q},-d(\bar{y})\bar{p},\bar{r},\bar{s}-d'(\bar{y})\bar{r}, \bar{t}-2d'(\bar{y})\bar{s} + (d'(\bar{y}))^2\bar{r}-d''\cdot\bar{p})$. Then $\bar{F}_{\bar{r}} = F_r - d'\cdot F_s + d'^2 F_t = \dfrac{1}{y_0'^2}[R\cdot y_0'^2 - S\, x_0' y_0' + T\cdot x_0'^2] \neq 0$.

Hence by theorem 55.1, there is a unique solution $\bar{z} = \bar{\varphi}(\bar{x},\bar{y})$ of

$$\bar{F}(\bar{x},\bar{y},\bar{z}, \frac{\partial\bar{z}}{\partial\bar{x}},\ldots) = 0$$

such that

$$\bar{\varphi}(0,\bar{y}) = g(\bar{y}); \quad \bar{\varphi}_{\bar{x}}(0,\bar{y}) = h(\bar{y})$$

Write $\varphi(x,y) = \bar{\varphi}(\bar{x}-d(\bar{y}),\bar{y})$. Then one can verify that this is a solution of (22.2). Also for $x = x_0(\mu)$, $y = y_0(\mu)$, $d(y_0(\mu)) = x_0(\mu)$ and $d'(y_0(\mu)) = \dfrac{x_0'(\mu)}{y_0'(\mu)}$ so that on the curve C,

$$\varphi(x_0(\mu), y_0(\mu)) = \bar{\varphi}(0, y_0(\mu)) = g(y_0(\mu)) = z_0(\mu).$$

$$\varphi_x(\qquad) = \bar{\varphi}_x(\qquad) = h(y_0(\mu)) = p_0(\mu)$$

$$\varphi_y(\qquad) = \bar{\varphi}_{\bar{x}}(\qquad)(-d'(y_0(\mu)) + \bar{\varphi}_{\bar{y}}(0, y_0(\mu)) = -h(y_0(\mu)\cdot\frac{x_0'(\mu)}{y_0'(\mu)} + g'(y_0(\mu)) = q\cdot(\mu)$$

COROLLARY 1: The quasi-linear equation (23.1) with analytic coefficients, and initial data as in (a) will have a unique local solution provided $a\cdot y_0'^2 - 2b x_0' y_0' + c x_0'^2 \neq 0$.

COROLLARY 2: For an arbitrary strip as defined in (a), if the equation is elliptic, there will always be a solution. (See section 25.)

THEOREM 55.3:

(a) Let $g(y)$ and $h(x)$ be two functions of class A^∞ for $|y-y_0| < \delta$ and $|x-x_0|<\eta_1$, respectively; suppose $g(y_0) = h(x_0)$. Set $z_0 = g(y_0)$, $p_0 = h'(x_0)$, $r_0 = h''(x_0)$, $q_0 = g'(y_0)$, $t = g''(y_0)$.

(b) Let $f(x,y,z,p,q,r,t)$ be of class A^∞ in a neighborhood of $(x_0,y_0,z_0,p_0,q_0,r_0,t_0)$. Then there exists a function $\varphi(x,y)$ such that:

(α) $\varphi(x,y)$ is of class A^∞ in a neighborhood R of (x_0,y_0): $|x-x_0| <\delta_1$, $|y-y_0| < \delta_2$.

(β) For every $(x,y)\varepsilon R$, $\varphi(x,y)$ satisfies the differential equation:

$$\frac{\partial^2 z}{\partial x \partial y} = f\left(x,y,z,\frac{\partial z}{\partial x},\frac{\partial z}{\partial y},\frac{\partial^2 z}{\partial x^2},\frac{\partial^2 z}{\partial y^2}\right) \tag{23.9}$$

(γ) For $|y-y_0| < \delta_1$, $\varphi(x_0,y) = g(y)$ and for $|x-x_0| < \delta_2$, $\varphi(x,y_0) = h(x)$.

The proof is again by the usual method of majorants. See Horn[1], p. 204. If $F_s \neq 0$, (54.1) can be reduced to (54.2).

Finally, we have to consider special kinds of equations which are linear, but cannot be classified as hyperbolic, elliptic or parabolic. M. Cibrario[5] in a series of papers considered equations of the form:

$$x^{2m}\frac{\partial^2 z}{\partial x^2} - \frac{\partial^2 z}{\partial y^2} = 0$$

$$y^{2k}\frac{\partial^2 z}{\partial x^2} - \frac{\partial^2 z}{\partial y^2} = 0$$

$$x^{2m}\frac{\partial^2 z}{\partial x^2} + \frac{\partial^2 z}{\partial y^2} = 0$$

which are of *mixed* elliptic-parabolic or hyperbolic-parabolic character since the coefficient of $\frac{\partial^2 z}{\partial x^2}$ may be 0. The solutions, quite long and involved, are given explicitly in terms of integrals. As representative of the type of result, we give:

THEOREM 55.4:

Let $\begin{cases} D_1 \\ D_2 \end{cases}$ be a finite region of the half plane $\begin{cases} x > 0 \\ x < 0 \end{cases}$ part of whose boundary is a segment $M \leq y \leq N$ of the y-axis and let $\begin{cases} \gamma_1 \\ \gamma_2 \end{cases}$ be the boundary of $\begin{cases} D_1 \\ D_2 \end{cases}$ except for the segment T: $M < y < N$. If continuous values are assigned along γ, then there exists in D_1 a unique solution of each equation of type:

$$h^{-2} x^{2(h+1)} \frac{\partial^2 z}{\partial x^2} + \frac{\partial^2 z}{\partial y^2} = 0 \qquad (h=1,2,3,\ldots)$$

which assumes these boundary values. If continuous values are assigned on $\gamma_1 + \gamma_2$, then there exists a unique solution of the above equation in the region: $D_1 + D_2 + D_3$ where D_3 is the set of all (x,y) with $x = 0$, $y \in T$, having these boundary values. (Under suitable conditions on γ_1 and assigned boundary values, the first solution has continuous first and second partial derivatives on $T \cdot D_1$. Under suitable conditions the second solution is analytic in D_1 and in D_2.)

Frankl[1] also considered equation of this sort - indeed, of a more general character:

$$y \frac{\partial^2 z}{\partial x^2} + \frac{\partial^2 z}{\partial y^2} + a(x,y)z_x + b(x,y)z_y + c(x,y)z = 0$$

where

$$z(x,0) = \bar{z}_0(x); \quad z_y(x,0) = \bar{q}_0(x).$$

Not existence theorems, but methods of construction of solutions from the characteristic integral strips of section 23, are found in many places. When the equation is of the Monge-Ampere type (23.5) then the problem reduces to finding a characteristic strip of first order (24.5) and Kamke[1] proves that the surface ψ " swept out" by the characteristic strip as the initial non-characteristic strip is traversed is a solution of the Cauchy problem, when ψ is of class A^2. Indeed, the solution of equations (24.5) can be replaced by another set, whose solution is obtained by the method of complete integrals.(Kamke[1], 390-398).

The results of the first two theorems of this section can be generalized to n dimensions, with no difficulty, and will not be repeated. For problem C, and linear equations, we give the following formulation due to F. John,[1] who established it as part of a discussion of nth order equations:

THEOREM 55.5: Let A_{ik}, A_i, f be regular analytic functions of real variables $(x_1, \ldots, x_n)$ which can be represented locally by power series. Let S be a surface, no normal to which has a characteristic direction- that is a direction whose direction numbers $(\alpha_1, \ldots, \alpha_n)$ satisfy $\Sigma A_{ik}\alpha_i\alpha_k = 0$, and let S be the image on an (n-1) dimensional sphere under a non-singular analytic transformation. Let prescribed values of u and of $\frac{\partial u}{\partial N}$ be given on S, which are regular analytic functions on S, $\frac{\partial u}{\partial N}$ being the normal derivative. Then for every closed subset Σ of S, there is a neighborhood in the $(x_1, \ldots, x_n)$ space in which there exists and is uniquely defined a solution of

$$(55.\quad) \sum_{i,k=1}^{n} A_{ik} \frac{\partial^2 u}{\partial x_i \partial x_k} + \sum_{i=1}^{n} A_i \frac{\partial u}{\partial x_i} + A_0 u = B$$

which takes on the prescribed values in Σ.

IV Partial Differential Equations of Order $n > 2$

56. The existence theorems for systems of first order differential equations of normal type given in Chapter II, section 18, were extended by Kowalewsky[1], Goursat, and others to systems of higher order as well. Indeed, the term *systems of Kowalewsky* is often applied to equations of the form:

$$(56.1)\quad \frac{\partial^{r_i} z_i}{\partial x^{r_i}} = f^i\left(x, y\ , \dots, y_n;\ z\ , \dots, z_m; \dots, \frac{\partial^{\alpha_0+\dots+\alpha_n} z_k}{\partial x^{\alpha_0} \partial y_1^{\alpha_1} \cdots \partial y_n^{\alpha_n}} \dots\right)$$

$$(i = 1, \dots, m)$$

where the terms on the right involve derivatives of orders up through and including those of order r_k of the unknown z_k, except that

$$\frac{\partial^{r_1} z_1}{\partial x^{r_1}}, \dots, \frac{\partial^{r_n} z_n}{\partial x^{r_n}}$$

do not enter any of the right hand members. The basic theorem follows:

THEOREM 56.1:

(a) Let $(x^0, y_1^0, \dots, y_n^{\ 0})$ be a given point and let $g^{ik}(y_1, \dots, y_n)$ $[i=1,\dots,m;\ k=0,1,\dots,r_i-1]$ be functions of class A^∞ in a neighborhood T_δ of this point. Set $z_i^0 = g^{io}(y_1^0, \dots, y_n^0)$ and $p_{i;\ \alpha_0, \alpha_1, \dots, \alpha_n}^{\ 0} =$

$$\left.\frac{\partial^{\alpha_1+\dots+\alpha_n} g^{i\alpha_0}}{\partial y_1^{\alpha_1}, \dots, \partial y_n^{\alpha_n}}\right|_{(y_1^0, \dots, y_n^0)} \equiv g^{i\alpha_0}_{\alpha_1, \dots, \alpha_n}(y_1^0, \dots, y_n^0)$$

for $[i=1,\dots,m;\ 0 \le \alpha_0+\dots+\alpha_n \le r_i;\ \alpha_0 \le r_i - 1]$.

(b_∞) Let $f^i(x, y_1, \dots, y_n;\ z_1, \dots, z_m, \dots\ p_{i;\ \alpha_0, \alpha_1, \dots, \alpha_n})$ be functions of class A^∞ $(i=1,\dots,m)$ in a neighborhood S_δ of

$(x^0, y_1^0, \ldots, y_n^0, z_1^0, \ldots, z_m^0, \ldots, p^o_{i;\ \alpha_0,\ldots,\alpha_n})$ $[i=1,\ldots,m;$ $0 \le \alpha_0+\ldots+\alpha_n \le r_i;\ \alpha_0 \le r_i - 1]$. Then there exists a unique set of functions $\overset{i}{\varphi}(x, y_1, \ldots, y_n)$ $(i=1,\ldots,m)$ such that

(α_0) $\overset{i}{\varphi}(x, y_1, \ldots, y_n)$ are of class A^∞ in a neighborhood R_{δ_1} of $(x^0, y_1^0, \ldots, y_n^0)$.

(β) $z_i = \overset{i}{\varphi}(x, y_1, \ldots, y_n)$ for $(i=1,\ldots,m)$ furnish a solution of (56.1) in R_{δ_1}.

(γ_0) $\overset{i}{\varphi}(x^0, y_1, \ldots, y_n) \equiv g(y_1, \ldots, y_n)$ for all $(y_1, \ldots, y_n)$ in the neighborhood T_{δ_1} of $(y_1^0, \ldots, y_n^0)$.

Proof: See Goursat[1], pp. 7-9. Same method of majorants used as that in proof of theorem 10.1 in Chapter II.

For a single mth order equation it would become:

THEOREM 56.2:

(a) Let $(x^0, y_1^0, \ldots, y_n^0)$ be a given point and $\overset{k}{g}(y_1, \ldots, y_n)$ $(k=0,1,\ldots,r-1)$ be functions of class A^∞ in a neighborhood T_δ of this point. Set $z^0 = g^0(y_1^0, \ldots, y_n^0)$ and $p^o_{\alpha_0,\alpha_1,\ldots,\alpha_n}\ldots)$

$$= \left.\frac{\partial^{\alpha_1+\ldots+\alpha_n} \overset{\alpha_0}{g}}{\partial y_1^{\alpha_1},\ldots,\partial y_n^{\alpha_n}}\right|_{(y_1^0,\ldots,y_n^o)} \equiv \overset{\alpha_0}{g}_{\alpha_1,\ldots,\alpha_n}(y_1^0, \ldots, y_n^0)$$

$[0 \le \alpha_0+\alpha_1+\ldots+\alpha_n \le r;\ \alpha_0 \le r-1]$

(b) Let $f(x, y_1, \ldots, y_n;\ z, \ldots, p_{\alpha_0,\alpha_1,\ldots,\alpha_n}\ldots)$ be a function of class A^∞ in a neighborhood S_δ of $(x^0, y_1^0, \ldots, y_n^0;\ z, \ldots, p^o_{\alpha_0,\ldots,\alpha_n})$. Then there exist a unique function $\varphi(x, y_1, \ldots, y_n)$ such that

(α) $\varphi(x,y_1,\ldots,y_n)$ is of class A^∞ in a neighborhood R_{δ_1} of $(x^o,y_1{}^o,\ldots,y_n{}^o)$.

(β_0) $z = \varphi(x,y_1,\ldots,y_n)$ is a solution in R_{δ_1} of:

$$(56.2)\quad \frac{\partial^r z}{\partial x^r} = f\left(x,y_1,\ldots,y_n;\ z;\ldots,\ \frac{\partial^{\alpha_o+\ldots+\alpha_n} z}{\partial x^{\alpha_o}\partial y_1{}^{\alpha_1},\ldots,\partial y_n{}^{\alpha_n}}\ldots\right)$$

$[0\le\alpha_0 + \alpha_1 +\ldots+\alpha_r;\ \alpha_o\le r-1]$

(γ_0) $\varphi_{x^k}(x^o,y_1,\ldots,y_n) \equiv g^k(y_1,\ldots,y_n)$ $(k=0,1,\ldots,r-1)$ in neighborhood T_{δ_1} of $(y_1{}^o,\ldots,y_n{}^o)$, where

$$\varphi_{x^k} \equiv \frac{\partial^k\varphi}{\partial x^k}.$$

COROLLARY

(a_∞) Let (x^o,y^o) be a given point and $g^k(y)$ $(k=0,1,\ldots,r-1)$ functions of class A^∞ in a neighborhood of this point. Set $z^o = g^o(y^o)$ and $q^o_{kj} = g^k_j(y^o) \equiv \frac{d^jg^k(y)}{dy^j}$ $(k=0,1,\ldots,r-1;\ j+k=0,\ldots,r-1)$

(b_∞) Let $f(x,y;\ z;\ q_{01},\ldots,q_{kj},\ldots,q_{r-1,1})$ be a function of class A^∞ in the neighborhood S_δ of $(x^o,y^o;z^o;\ldots,q^o_{kj}\ldots)$. Then there exists a unique function $\varphi(x,y)$ such that

(α_∞) $\varphi(x,y)$ is of class A^∞ in a neighborhood R_{δ_1} of (x^o,y^o)

(β) $z = \varphi(x,y)$ is a solution in R_{δ_1} of

$$(56.3)\quad \frac{\partial^r z}{\partial x^r} = f\left(x,y;\ z;\ldots,\ \frac{\partial^{j+h} z}{\partial x^k\partial y^j}\ldots\right)$$

(γ_0) $\varphi_{x^k}(x^o,y^o) \equiv g^k(y)$ $(k=0,1,\ldots,r-1)$

Completely integrable systems (see section 20, page 79) can also be studied in more variables, as was done by König, Riquier, and others. In particular, analogue of theorem 56.2 holds for *systems of Konig* (Goursat[1], p.41).

57. Beginning with the work of Riquier and Meray in 1892, the question of when a system of m partial differential equations in n independent variables and r unknown functions:

$$F_k(x_1,\dots,x_n, z_1,\dots,z_r,\dots \frac{\partial^{\alpha_1+\dots+\alpha_n} z_i}{\partial x_1^{\alpha_1}\cdots\partial x_n^{\alpha_n}},\dots) = 0$$

is capable of a solution, how many arbitrary parameters or functions this "general solution" depends on, and what additional conditions are necessary to obtain a unique solution, has been considered by many people. Riquier[1], Janet[1], J. M. Thomas[1], T. Y. Thomas and E. Titt[1] have done fundamental work on this question -- in general, what is done is to determine conditions under which the given system can be solved to yield an "orthonomic" system, and then conditions under which the successive differentiation of the "principal derivatives" will yield the same expression, regardless of the order of differentiation. Indeed, Janet[1] (p. 23) discusses a regular procedure by which, after a finite number of operations, either the impossibility of the solution of the proposed system is shown, or else one is lead to a canonical form for which one knows a precise existence theorem. Only local, analytic solutions are considered, and analytic equations and initial data; and the terms in which the conditions are given are too special to be given here, except for the cases already mentioned in chapters 2 and 3. Recently, Coutrez[1] gave a different form of the conditions that are necessary to supplement the Cauchy data, when it is given on a characteristic surface, (where there is no unique solution, determined by the data alone). Lednev[1] and Cramlet[1] also considered this matter in a manner quite different from Riquier's. In a book I have not seen, B. Levi[1] summarized many of the fundamental theorems for analytic questions of this type.

58. The quasi-linear equation of order r in n independent variables is:

$$(58.1)\quad \sum_{i_1+\ldots+i_n=r} a_{i_1,\ldots i_n}(x_1,\ldots x_n)\frac{\partial^r u}{\partial x_1^{i_1}\ldots\partial x_n^{i_n}} + g(x_1,\ldots,x_n,\ldots)=0$$

where g involves derivatives of order less than r. When $a_{r,0,\ldots,0} \neq 0$, and all a's and f are of class A^{∞}, theorem 56.2 will insure the existence of a local solution to problem I. For non-analytic coefficients, we have results for n = 2. (See Courant-Hilbert, 138-141 for terminology of characteristics as applied to equations (58.1) - no existence theorems) Here the equation is

$$(58.2)\quad \sum_{k=0}^{r} a_{k,r-k}(x,y)\frac{\partial^r u}{\partial x^k \partial y^{r-k}} + g(x,y,z,\ldots) = 0$$

As in the two variable case, we call

$$(58.3)\quad \sum_{k=0}^{r} a_{k,r-k}(x,y)\cdot(-x')^{r-k}(y')^k = 0$$

the characteristic equation of (24.2). At a particular point P^0, if the characteristic form $\sum a_{k,r-k}\,\alpha_1{}^k\alpha_2{}^{r-k}$ has r real linear factors, then the equation is called hyperbolic and the characteristic equation can be written

$$(58.4)\quad \prod_{j=1}^{n}(\eta_j\, y' - \xi_j\, x') = 0$$

Each solution of this equation through P^0 is called a curve with a characteristic direction at P^0; there are at most r such directions. If the equation is hyperbolic in R, - that is, at every point of R - then the solutions of (58.4)

$$y = \varphi_j(\mu; x_0, y_0);\quad x = \psi_j(\mu; x_0, y_0)$$

are called the characteristic curves of (58.2). If all the factors are distinct, the equation is called totally hyperbolic in R. If $\sum a_{k,r-k}\alpha_1{}^k\alpha_2{}^{r-k}$ is a definite form, so that it has no real linear factors at P^0 or in a region R, then the equation is called elliptic at P^0 or in R. Finally if in the factored form there appear less than r factors, then the equation is called parabolic at P^0 or in R. (Sometimes, corresponding to (58.4), the equation (58.2) is found in the literature in the symbolic form $\prod_{j=1}^{n}(\eta_j\frac{\partial}{\partial x} - \xi_j\frac{\partial}{\partial y})u+g = 0.$) If it is known that $a_{ro} \neq 0$ throughout R, then each η_j can be taken as 1 in (58.4).

When the coefficients $a_{k,r-k}$ are all constants, then the characteristic curves are the straight lines:

$$\eta_j\, y - \xi_j\, x = \text{const.}$$

the particular curves through any point (x_0, y_0) being:

$$\eta_j (y - y_0) = \xi_j (x - x_0)$$

If $r = s+t$, and all $A_{k,r-k} = 0$ except $A_{st} \neq 0$, then we can write the equation as:

$$(58.5)\quad \frac{\partial^{s+t} z}{\partial x^s \partial y^t} = f(x, y, z, \ldots, z_{ik}, \ldots)$$

The characteristic equation becomes:

$$(-x')^t (y')^s = 0$$

so the characteristic curves are: $x = \text{const.}$ and $y = \text{const.}$ When $r > 2$, this equation is not totally hyperbolic, since the characteristic roots are not distinct, but Zwirner[1] proved the following analogues of the Goursat problem:

THEOREM 58.1: Let f be a continuous function with $|f(x,y,0,\ldots,0)| \leq \frac{1-K}{N}(k-1)$ for (x,y) in $\overline{R}$: $a \leq x \leq b$, $c \leq y \leq d$ and let f satisfy a Lipschitz condition:

$$|f(x,y,z,\ldots) - f(x,y,\overline{z},\ldots)| \leq L \sum_{i,j} |z_{ij} - \overline{z}_{ij}|$$

where

$$L \leq K \sum_{i,j} \frac{(b-a)^{s-i}}{(s-i)!} \frac{(d-c)^{t-j}}{(t-j)!}$$

Let the following two systems of characteristics in $\overline{R}$ be assigned:

$$x = x_i \ (i=1,\ldots s) \quad \text{for } a \leq x \leq b\ ; \qquad y = y_j \ (j=1,\ldots t) \text{ for } c \leq y \leq d$$

Then there exists a unique function $\varphi(x,y)$ in R such that $z = \varphi(x,y)$ satisfies (58.5) in R and such that:

$$\varphi(x_i, y) = 0 \quad \text{for } c \leq y \leq d; \qquad \varphi(x, y_j) = 0 \text{ for } a \leq x \leq b.$$

The proof is based on a functional theorem due to Graves and Hildebrandt, of the same character as Schauder's theorem. In an earlier paper, he established the existence of a solution such that $\varphi(x_0, y)$ and its first s-1 derivatives assumed prescribed values on $x = x_0$ while $\varphi(x, y_0)$ and its first t-1 derivatives assumed given values on $y = y_0$.

Mangeron[1] considered a special case of (58.5), where s=t=2, obtaining a result like that of the above theorem, but using the method of successive approximation beginning, as in section B of chapter 3, with the solution of

$$\frac{\partial^4 z}{\partial x^2 \partial y^2} = \varphi(x,y)$$

F. Bureau, in a succession of papers, the last of which concerns:

$$f\left(\frac{\partial}{\partial x}, \frac{\partial}{\partial y}, \frac{\partial}{\partial z}, \ldots, \frac{\partial}{\partial t}\right)u = g(x,y,z,\ldots,t)$$

where f is a form of degree n, with constant coefficients, obtains solutions of the problem of Cauchy for totally hyperbolic equations, given the initial data on the plane t = 0. He does this by extending the method of Hadamard, used for second order linear equations, first to a simple fourth order case and then generalizes it.

Rosenblatt[1] has considered solutions of:

$$\Delta\Delta u = F(x,y;u,u_x,u_y,u_{xx},u_{xy},u_{yy})$$

where a Lipschitz condition holds for F; Picard successive approximations used. He also considered (Rosenblatt[2])

$$(58.6) \qquad \Delta_{2m} u = F\left(x,y,u,\ldots, \frac{\partial^{2m-2} u}{\partial y^{2,-2}}\right).$$

where the left side is: $\left(\frac{\partial^2}{\partial x^2} + \frac{\partial^2}{\partial y^2}\right)^m$, so that the characteristics are solutions of:

$$(y'^2 + x'^2)^m = 0$$

that is,

$$y \pm ix = \text{constant}$$

Then if the values of u and its normal derivatives up to order m-1 are given on the circumference C of a circle with center at the origin and radius R, he proved that there exists a solution of (58.6) inside the circle which approaches these values as u approaches the boundary, provided F is continuous in K+C and satisfies a certain generalized Lipschitz condition there, and certain inequalities of Holder type hold.

Winants[1] showed that if $f(x,y,z,p,q,s)$ is a function of class A^0 in

$$D: \begin{cases} 0 \le x \le \alpha; & |z| \le H; \ |s| \le S \\ 0 \le y \le \alpha; & |p| \le P; \ |q| \le Q \end{cases}$$

and if f satisfies a Lipschitz condition

$$|f(x,y,z_1,p_1,q_1,s_1) - f(x,y,z_2,p_2,q_2,s_2)| \le k_1 |z_1-z_2|$$
$$+ k_2 |p_1-p_2| + k_3 |q_1-q_2| + k_4 |s_1-s_2|$$

then there exists a function $\varphi(x,y)$ such that $z = \varphi(x,y)$ is a solution of:

$$\frac{\partial^3 z}{\partial x^2 \partial y} - \frac{\partial^3 z}{\partial x \partial y^2} = f(x,y,z,\frac{\partial z}{\partial x},\frac{\partial z}{\partial y},\frac{\partial^2 z}{\partial x \partial y})$$

and such that

$$z = \frac{\partial z}{\partial x} = \frac{\partial^2 z}{\partial x^2} = 0 \text{ for } x = y.$$

(By Picard method of successive approximation.)

Winants[1] also found a solution of the same equation, under the same hypotheses, with the boundary conditions:

$$(58.7) \quad \begin{cases} \varphi(x,0) = \varphi_1(x) \\ \varphi(0,y) = \varphi_2(y) \\ \varphi_x(0,y) = \lambda(y) \end{cases}$$

where $\varphi_1(0) = \varphi_2(0)$ and $\varphi_1'(0) = \lambda(0)$, provided $\varphi_1, \varphi_2, \lambda$ are of class A^2.

59. The linear equation in n variables was considered in detail by Gevrey[1]:

$$(59.1) \quad \sum_{k=0}^{r} \sum a_{i,\dots i_n}(x_1,\dots,x_n) \frac{\partial^k u}{\partial x_1^{i_1}\dots\partial x_n^{i_n}} = f(x_1,\dots,x_n)$$

when the coefficients were analytic; he obtained a solution to the problem of Cauchy, when the data was analytic. F. John[1] considered question of uniqueness in a paper where he stated the solution of the Cauchy problem as follows:

Let S be a surface which is the image of an open (n-1) dimensional full sphere under a non-singular analytic transformation, and let no normal to S have a characteristic direction -- one whose direction numbers $(\alpha_1,\ldots,\alpha_n)$ satisfy

$$\Sigma\, a_{i_1,\ldots,i_n}\, \alpha_1^{\,i_1} \ldots \alpha_n^{\,i_n} = 0 \qquad (i_1+\ldots+i_n=r)$$

Let the values of u, $\frac{\partial u}{\partial N},\ldots,\frac{\partial^{r-1} u}{\partial N^{r-1}}$ be prescribed on S, where $\frac{\partial u}{\partial N}$ is the normal derivative. And let these values, as well as the coefficients of the equation, be analytic. Then for any closed subset Σ of S there is a neighborhood of Σ in the $(x_1,\ldots,x_n)$ plane in which there exists and is uniquely determined a regular analytic solution of (59.1) assuming the prescribed initial values on S.

Picone[1], in a paper I have not seen, considered the same equation, and by using a Green's formula and an operator, reduces the problem to one which can be solved by successive approximation.

For the important equation:

$$(59.2) \quad \alpha\, \Delta\, \Delta\, u + 2\, B\, \Delta\, u + \gamma\, u = 0$$

Bremenkamp[1] proved the following theorem for the analogue of the Dirichlet problem:

THEOREM 59.1: Let C be a closed curve without double points and a definite tangent at each point; let u and either Δu or $\frac{\partial u}{\partial n}$ be given at each point on C. Let $\alpha\gamma-\beta^2$ have the same sign at all interior points of C. Then there exists a unique function u satisfying (59.2) in the interior of C and assuming assigned values on the boundary.

Other special third and fourth order equations have been considered; the hypotheses vary, but usually the problem is reduced to one of those previously solved, or else a straightforward solution is obtained by the method of successive approximation. Germay[4] extended the method of Riemann to special kinds of fourth order systems which are generalizations of second order hyperbolic systems. Théodoresco[1] **also** took special kinds of homogeneous **fourth** order linear equations and found *elementary solutions*. F. Bureau[1] considered the problem of Cauchy for

$$D_\mu D_\psi u = F(x_1, x_2, x_3),$$

where $D_\varphi = a_3 \frac{\partial^2}{\partial x_3^2} - a_1 \frac{\partial^2}{\partial x_1^2} - a_2 \frac{\partial^2}{\partial x_2^2}$, ($a_i, b_i$ constant;

$D_\psi = b_3 \frac{\partial^2}{\partial x_3^2} - b_1 \frac{\partial^2}{\partial x_1^2} - b_2 \frac{\partial^2}{\partial x_2^2}$ $\quad 0 < a_3 < a_2 < a_1,\ 0 < b_1 < b_2 < b_3$).

A special kind of linear equation which can be attacked by the same means as those used for parabolic equations (see papers by Rothe cited in III, G) is

$$(59.3) \quad \sum_{i=0}^{n} p_i(x) \frac{\partial^{n-i} u}{\partial x^{n-i}} = Au_t + Bu_{tt}$$

with initial conditions:

$$(59.4) \quad u(x,0) = \alpha(x);\ u_t(x,0) = \beta(x) \quad (a \leq x \leq b)$$

and boundary conditions:

$$(59.5) \quad L_i(u) = 0 \quad (t \geq 0)$$

where L_i is a linear form in $u(a,t)$, $u_t(a,t), \ldots, u_{t^{n-1}}(a,t)$, $u(b,t), \ldots, u_{t^{n-1}}(b,t)$. When $u = X(x)\,T(t)$ is substituted in (59.3) the resulting equation splits into two ordinary equations; (setting each factor = λ first); then the Sturm-Liouville problem for the nth order ordinary ewuation in x is solved. See, for example, Tamarkin and Feller[1], pp. 65 ff. (In the same book, pp. 58-60, there are statements concerning equations similar to (59.3) except that the right hand side is f(x,t).) The conditions for solution are reducible to conditions that the Sturm-Liouville problem have a solution. (L(u) must be self-adjoint, etc.)

S. Bruk[1] considered parabolic systems of equations:

$$\frac{\partial u_i}{\partial t} = \Sigma\, A_{ij}^{(M)}(t)\, \frac{\partial^M u_j}{\partial x_1^{m_1} \ldots \partial x_n^{m_n}} \quad \Sigma\, B_{ij}(t,x)\, \frac{\partial^k u_j}{\partial x_1^{k_1} \ldots \partial x_n^{k_n}} - + f_i(t,x) \quad (i=1,..n)$$

he reduced the problem to a system of intregro-differential equations which he solved by successive approximations; he also obtained fundamental solutions.

60. The linear equation with constant coefficients can be written in one of two ways:

$$(60.1) \quad \sum_{r=0}^{p} \sum_{s=0}^{q} A_{rs} \frac{\partial^{p+q-r-s} z}{\partial x^{p-r} \partial y^{q-s}} = f(x,y)$$

$$(60.2) \quad \sum_{k=0}^{n} a_{nk} \frac{\partial^n z}{\partial x^k \partial y^{n-k}} + \sum_{k=0}^{n-1} a_{n-1,k} \frac{\partial^{n-1} z}{\partial x^k \partial y^{n-1-k}} + \ldots + a_{00} z = f(x,y)$$

The second form enables one to see more readily the characteristic form:

$$P(\alpha_1,\alpha_2) = \sum_{k=0}^{n} a_{nk}\, \alpha_1^k \alpha_2^{n-k}$$

By section 58, if this form can be factored into linear factors (real or complex)

$$P(\alpha_1,\alpha_2) = \prod_{k=1}^{n} (\eta_k\, \alpha_1 + \xi_k\, \alpha_2)$$

then the characteristic curves through any point (x_0,y_0) are the straight lines

$$\eta_k\, (y-y_0) = \xi_k(x-x_0)$$

or

$$\eta_k\, y - \xi_k x = \text{constant}$$

If there are $b>2$ characteristics through any point, they are said to be excessive in number; if $b = 2$, they are sufficient; if $b<2$, they are deficient. (These terms can be used for any number of independent variables). When $b=2$, the transformation

$$\bar{x} = \eta_1 y - \xi_1 x \quad , \quad \bar{y} = \eta_2 y - \xi_2 x \qquad (\eta_1 \neq 0)$$

will reduce (60.2) to (60.1) with $A_{00} \neq 0$. (If we started with 60.1, with $A_{00} = 0$, same method could be used.) Then, the characteristics are just:

$$\bar{x} = \text{constant}, \quad \bar{y} = \text{constant}$$

For the particular case of the homogeneous equation

$$(60.3) \quad \sum_{k=0}^{n} a_k \frac{\partial^n z}{\partial x^k \partial y^{n-k}} = 0$$

MacDuffee[1] found a recursion formula for getting all of the polynomial solutions. One can verify that a solution of equation (60.3) will always be of the form:

$$(60.3a) \quad z = \prod_{k=1}^{n} F_k(\eta_k y - \xi_k x)$$

where $F_k(t)$ is a function of class A^1. Indeed, if the characteristics are distinct, this is the general solution, and can be used to solve any particular boundary problem for this equation.

T. Chaundy[1] has given a theory of the general equation (60.2) when $f(x,y) = 0$. As we have seen, this can be reduced to (60.1) with non-zero leading term, provided there are exactly 2 characteristics through each point. That is, for b=2, he considered:

$$(60.4) \quad \sum_{r=0}^{p} \sum_{s=0}^{q} A_{rs} \frac{\partial^{p+q-r-s} z}{\partial x^{p-r} \partial y^{q-s}} = 0 \text{ where } A_{00} = +1.$$

Since this is a generalization of Riemann's equation

$$\frac{\partial^2 z}{\partial x \partial y} + a\frac{\partial z}{\partial x} + b\frac{\partial z}{\partial y} + cz = 0$$

it can be studied by analogous methods. The adjoint equation would be merely:

$$(60.5) \quad \sum_{r=0}^{p} \sum_{s=0}^{q} A_{rs} (-1)^{p+q-r-s} \frac{\partial^{p+q-r-s} z}{\partial x^{p-r} \partial y^{q-s}} = 0$$

so that if $z = V(x,y)$ is a solution of (60.5), $z = V(-x,-y)$ is a solution of (60.4), or more generally $V(X-x, Y-y)$ where X and Y are constants.

The *fundamental solution* $U(x,y;X,Y)$ is a solution of (60.5) when (x,y)

are the variables and (X,Y) the parameters, a solution of (60.4) when (X,Y) are the variables and (x,y) the parameters, and which has certain values on characteristics x = X, y = Y. Define:

$$\phi(x,y) \equiv \Sigma \frac{T!\, x^{R+p-1} y^{S+q-1}}{(R+p-1)!(S+q-1)!} \prod_{r,s}{}' \frac{A_{rs}^{t(r,s)}}{t(r,s)!}$$

where Σ is taken as a multiple sum over all negative integral values of every t(r,s), Π′ omits the term (0,0) and

$$R = \sum_{r,s}{}' rt(r,s),\quad S = \sum_{r,s}{}' st(r,s),\quad T = \sum_{r,s}{}' t(r,s)$$

where again the term (0,0) is omitted in sums. Then $\phi(0,x) = \phi(0,y) = 0$, and indeed $\phi_{m,0}(0,y) = 0$ for $m \leq p-2$ while $h(y) = \phi_{p-1,0}(0,y)$ satisfies:

$$(60.6)\qquad \sum_{s=0}^{q} A_{0s} \frac{d^q h}{dy^{q-s}} = 0$$

$\phi_{0,n}(x,0) = 0$ for $n \leq q-2$ and $\phi_{0,q-1}(x,0)$ satisfies an analogous equation to (60.6) and $\phi_{m,n}(0,0) = 0$ for $m \leq p-1$, $n \leq q-1$ unless $m = p-1$ and $q = n-1$ whence $\phi_{p-1,q-1}(0,0) = 1$. Then we define the *fundamental solution*

$$U(X,Y{:}x,y) \equiv \varphi(X-x,Y-y).$$

We now can state the following result:

THEOREM 60.1:

Let $g_n(x)$ $(n=0,\ldots,q)$ and $h_\lambda(y)$ $(\lambda=0,\ldots,p)$ be given functions of class A^q and A^p respectively. Let (ξ,η) be given numbers and $g_n^{(m)}(\xi) = h_m^{(n)}(\eta)$. Then there exists a function $\varphi(x,y)$ which satisfies the differential equation (60.4) and such that:

$$\varphi_{0,n}(x,\eta) = g_n(x) \text{ and } \varphi_{\lambda,0}(\xi,y) = h_\lambda(y)$$

$$(n=0,\ldots,p-1,\ \lambda=0,\ldots,q-1)$$

It is given by the formula:

$$\varphi(x,y) = \int_{\xi}^{x} \sum_{r=0}^{p} \sum_{s=0}^{q-1} (-A_{rs}) \sum_{n=0}^{q-s-1} \frac{d^{p-r} g_n(t)}{dt^{p-r}}$$

$$\phi_{0,q-s-n-1}(x-t, y-\eta)\, dt + \int_{\eta}^{y} \sum_{r=0}^{p-1} \sum_{s=0}^{q} (-A_{rs})$$

$$\sum_{m=0}^{p-r-1} \frac{d^{q-s} h_m(t)}{dt^{q-s}} \phi_{p-r-m-1,0}(x-\xi, y-t)\, dt$$

$$+ \sum_{r=0}^{p-1} \sum_{s=0}^{q-1} (-A_{rs}) \sum_{m=0}^{p-r-1} \sum_{n=0}^{q-s-1} \frac{d^m g_n(\xi)}{dt^m}$$

$$\phi_{p-r-m-1,q-s-m-1}(x-\xi, y-\eta)$$

In the above formulas, the prescribed values of φ and its derivatives are given along constants $x = \xi$, $y = \eta$ through (ξ,η). If the prescribed data is given on an arbitrary curve C, which is intersected by the characteristic $\begin{cases} x = X \\ y = Y \end{cases}$ in the single point $\begin{cases} (X,\eta) \\ (\xi,Y) \end{cases}$ then if we write $\varphi(x,y) = F(X,\eta) + G(\xi,Y)$, the solution will be

$$(60.7) \quad \varphi(x,y) = \tfrac{1}{2} \int_{(\xi,Y)}^{(X,\eta)} [(F_x - G_x)\, dx + (F_y - G_y)\, dy]$$

$$+ \tfrac{1}{2} [F(X,\eta) + G(X,\eta) + F(\xi,Y) + G(\xi,Y)]$$

where integration is along C, and F, G, F_x, G_x are given in terms of the fundamental solution and prescribed values of φ on C.

In a second paper, Chaundy[2] discusses the case in which there are *excessive* characteristics. $(b>2)$

Various special linear differential equations with constant coefficients have been considered. Problems analogous to Dirichlet and Neumann problems can be solved for the biharmonic equation

$$\Delta\Delta u = 0 \quad \left(\Delta = \frac{\partial^2}{\partial x^2} + \frac{\partial^2}{\partial y^2}\right)$$

Schroder[1] gave a comprehensive discussion of both the " exterior" and " interior" problems in 2 and 3 dimensions, establishing existence and uniqueness of functions which are solutions inside a region and take on prescribed values, together with first derivatives, on the boundary, under rather special restrictions on the type of boundary that can be used. His method was to replace the single equation by a system of equations and then use integral equations. Bergmann[1] reduced the problem to a consideration of $\Delta u = 0$, by letting $\varphi(z,\bar{z}) = \varphi(x,y)$ where $x+iy = z$, $x-iy = \bar{z}$ and considering the corresponding equation in z and $\bar{z}$. Aronsajn and Weinstein[1] considered $\Delta\Delta u + \lambda u = 0$, which they solved by using a succession of " intermediate'' problems. Difference equations have been applied by Waschakidze[1] and Courant, Friedrichs, and Lewy[1]. Southwell[1] and others used his relaxation methods. A. E. Green[1] used an expansion in double Fourier series to get a solution in the inside of a rectangle which assumed, with its first normal derivatives, the values 0 on the boundary. (This does not exhaust the papers on this equation, but gives some idea as to methods.)

The equation

$$\Delta^k u \equiv \sum_{j=0}^{k} \frac{k!}{j!(k-j)!} \frac{\partial^{2k} u}{\partial x^{2j} \partial y^{2k-2j}} = 0$$

has, as was shown in section 58, only two distinct families of characteristics - the imaginary lines:

$$y \pm ix = \text{const.}$$

Hence although this is of the form (60.3a) the solution $F(y-ix) + G(y+ix)$ is not the general solution. An existence theorem does follow from Rosenblatt's theorem, given in section 58 - indeed in its proof, he began with this case. Bremenkamp[1] proved first the uniqueness, and then the existence of a solution of $\Delta^k u = 0$ inside a closed curve C, if u and its first k-1 normal derivatives are prescribed on the boundary, provided that certain "regularity" conditions hold, and the boundary is of a certain kind. He also showed that such a solution can be written as a sum of solutions of the corresponding problem for k=1.

Vecoua[1] considered the equation

$$\Delta^k u + \sum_{i=0}^{k-1} a_{k-i}\, \Delta^i u = 0$$

and obtained an elementary solution by reducing the problem to a simpler one for $u(z,\bar{z})$ as did Bergmann above. He also showed that all real roots of this equation can be represented: $u = \Sigma A_n f_n(r)\phi_n(\theta)$ where each $f_n(r)\phi_n(\theta)$ is a particular solution.

Cimino[1] found an explicit formula, in terms of spherical harmonics, and then established the existence of the solution of the Cauchy problem for $\Delta^m u = 0$ (where $\Delta u = u_{xx}+u_{yy}+u_{zz}$, and then is transformed into spherical coordinates ρ, ϕ, θ) in the interior of a sphere, given the Cauchy initial data on the sphere $\rho = r$. Columbo[1] solved a problem of Goursat for:

$$\left[\frac{\partial}{\partial x} + \lambda \frac{\partial}{\partial y}\right] \left[\frac{\partial}{\partial x} + \frac{\partial}{\partial y}\right] \frac{\partial^2 z}{\partial x \partial y} = 0.$$

For third order equations, Sjostrand[1] solved various problems for

$$\frac{\partial}{\partial x}\left(\frac{\partial^2}{\partial x^2} + \frac{\partial^2}{\partial y^2}\right) u = 0.$$

Winants considered, in a series of papers[3], the problem of Cauchy for

$$\frac{\partial^3 z}{\partial x^2 \partial y} - \frac{\partial^3 z}{\partial x \partial y^2} = 0$$

and for

$$A \frac{\partial^3 z}{\partial x^3} + B \frac{\partial^3 z}{\partial x^2 \partial y} + C \frac{\partial^3 z}{\partial x \partial y^2} + D \frac{\partial^3 z}{\partial y^3} = f(x,y).$$

Rellich[1] considered solution of

$$\frac{\partial^3 u}{\partial x^2 \partial t} - au = f(x,t)$$

such that

$$u(x,0) = \omega(x) \qquad (0 \le x \le 1)$$

$$u(0,t) = \varphi(t) \qquad (t \ge 0)$$

$$u(1,t) = \omega(t) \qquad (t \ge 0)$$

(where $\omega(0) = \psi(0)$ and $\omega(1) = \varphi(0)$) by setting up a *fundamental solution* $U(x,t;\xi,\tau)$ and then writing

$$U(x,t) = \int_0^t \int_0^x f(\xi,\tau)\; U(x,t;\xi,\tau)\; d\tau\, d\xi .$$

The solution exists if $f(x,t)$ is continuous in $\begin{matrix} 0 \le x \le 1; \\ 0 \le t \end{matrix}$ $\omega(x)$ is of class A^2 on $0 \le x \le 1$, and $\varphi(t)$ and $\psi(t)$ are continuous for $0 \le t$.

61. The general equation in 2 variables of order r is:

$$(61.1) \qquad F(x,y,z, \frac{\partial z}{\partial x}, \ldots \frac{\partial^{i+k} z}{\partial x^i \partial y^k}, \ldots) = 0$$

where $F(x,y,z,q_{10},\ldots,q_{jk},\ldots,q_{ro},\ldots,q_{or})$ is a function of class A^1 in U. Friedrichs and Lewy[1] called such an equation totally hyperbolic if, setting

$F_j = \dfrac{\partial F}{\partial q_{r-j,j}}$,

$$\Delta_r(\rho) = \begin{vmatrix} F_0 & F_1 & \cdots & F_n \\ 1 & \rho & 0 \; . \; . & 0 \\ 0 & 1 & \rho \; . \; . & 0 \\ . & . & . \; . \; . & . \\ 0 & . & . \; . \; 1 & \rho \end{vmatrix} = 0$$

has exactly n real roots at all points of the region U. Then the characteristic equation

$$(61.2)\quad \begin{vmatrix} F_0 & F_1 & \cdots & F_n \\ x' & y' & \cdots & 0 \\ 0 & x' & y' \cdots & 0 \\ \cdot & \cdot & \cdots & \cdot \\ 0 & \cdot & \cdots x' & y' \end{vmatrix} \equiv \sum_{k=0}^{r} F_{r-k} (y')^k (-x')^{r-k} = 0$$

is satisfied by exactly n values of y'/x', which are called characteristic directions. In the neighborhood of any point, a linear transformation can be made so that neither $x = 0$ nor $y = 0$ is a characteristic direction, and hence both F_0 and F_n will be different from 0.

By reducing the solution of the problem to the solution of n first order differential equations, they claim to have proved the following: (I think the theorem is correct, but as has been pointed out, the proof is incorrect.)

Let C_0: $x = x(t)$, $y = y(t)$ be a curve which is nowhere tangent to a characteristic direction and let $z(t)$, $q_{10},(t),\ldots,q_{ik}(t),\ldots$ $q_{or}(t)$, be given so that all of these have bounded third derivatives; let F have bounded derivatives of order 4 in a region U containing S: $x = x(t)$, $y = y(t)$, $z(t),\ldots q_{ik}(t),\ldots$ $[t\varepsilon T]$, and suppose that S is an integral strip; that is, for each point P of S,

$$F(x(t),y(t),\ldots,q_{ik}(t),\ldots) \equiv 0$$

$$F_0'(t) = F_{0,x}\cdot x'(t) + F_{0,y}\, y'(t) \qquad (F_j'(t) = F_{x^i y^{r-j}})$$

$$F_1'(t) = F_{1,x}\cdot x'(t) + F_{1,y}\cdot y'(t)$$

$$\cdot \qquad \cdot \qquad \cdot \qquad \cdot$$

$$F'_{m-1}(t) = F_{n-1,x}\, x'(t) + F_{n-1,y}\, y'(t)$$

Then there exists a unique function $\varphi(x,y)$ which is of class A^{r+2} in a region R containing C_0, which satisfies (61.1) in R, and such that along C_0

$$\varphi_{x^i y^k}(x(t),y(t)) = q_{ik}(t)$$

Furthermore, the values of $\varphi(x,y)$ at a point P depend only on the values of S on the part of C contained between two (external) characteristics thru P.

Cinquini-Cibrario[5] considered the same equation, but she began by setting up the equations for a characteristic strip. Let $\Delta_r(\rho) = 0$ have n distinct real roots $\rho_1,\ldots,\rho_n$. If S: $z=\varphi(x,y)$ were an integral surface, each equation: $\frac{dy}{dx} = \rho_i$ would determine a curve $y=y(x)$ which one could call a characteristic curve on S. But then $y(x)$, $z(x)$,... would need to satisfy a certain set Σ of ordinary differential equations, as in the one variable case. Then these equations, regardless of whether there is an integral surface, are called the characteristic strip equations, and any solution of them a characteristic strip. One can show that F is constant along such a strip, so that if F = 0 at an initial value, it is 0 along the strip. For a given initial integral strip which is non-characteristic, a unique integral surface is determined. When the given strip is characteristic, there are infinitely many integral surfaces containing it, provided the functions $y(x),z(x),\ldots,q_{ik}(x),\ldots$ are of class A^2 with second derivatives satisfying Lipschitz conditions. The existence of the solution of the system on which this depends is shown in another paper .

Lednev[1] considered the solution of a generalized Cauchy problem for systems (56.1), relaxing somewhat the requirements of analyticity - f^i and φ^i need only be continuous in some variables.

BIBLIOGRAPHY

Akushsky, I. [1] *Numerical solution of the Dirichlet equation with the aid of perforated card machines.* Acad. Sci. U.R.S.S., C. R. (Doklady) (N.S.) 52, 375-378 (1946)

Aronszajn, N. and Weinstein, A. [1] *On the unified theory of eigenvalues of plates and membranes.* Amer. J. Math. 64, 623-645 (1942)

Badescu, R. [1] *Simplificazioni ed estrinzioni de metode di Picone---.* R. Accad. Naz. Lincei, Cl. Sc. Fis. Mat. Nat., Rend., (6) 27, 624-630 (1938)

Baiada, E. [1] *Sul teoreme d'esistenza per de equazioni alle derivate parziale del primo ordine.* Scuola Norm. Sup. Pisa, Ann. (2) 12, 135-145 (1943)

Bergmann, S. [1] *The approximation of functions satisfying a linear partial differential equation.* Duke Math. J. 6, 537-561 (1940)

Bernstein, S. [1] *Sur l'intégration aux dérivées partielles du type elliptique.* Math. Ann. 69, 82-136 (1910); ibid, 96, 633-647 (1926)

Bourgin, D. G. and Duffin, R. [1] *The Dirichlet problem for the vibrating string equation.* Amer. Math. Soc., Bull. 45, 851-858 (1939)

Bottema, O. and Bremekamp, H. [1] *On the solutions of the equation ... which satisfy certain boundary conditions.* K. Akad. Wetensch., Afdeel. nat., Proc. 49, 424-443 - Indagationes Math. 8, 279-298 (1946).

Bremekamp, H. [1] *Sur l'existence et la construction des solutions des équations aux dérivées partielles du quatrieme ordre.* K. Akad. Wetensch., Afdeel. nat., Proc. 45, 675-680 (1942)

[2] *On the existence of a solution of ----.* ibid, 49, 185-193 and 302-318 (1936)

Bruk, S. Z. [1] *The fundamental solutions of a system of differential equations of parabolic type.* Akad. Nauk S.S.S.R., Doklady (N.S.) 60, 9-12 (1948)

Bureau, F. [1] *Sur le probleme de Cauchy pour les équations aux dérivées partielles, totalement hyperboliques, d'ordre plus grand que 2me.* Acad. Roy. Belgique, Cl. Sci., Bull. (5) 33, 587-610 (1947)

[2] *L'intégration des équations linéaires aux dérivées partielles--.* Soc. Roy. Sci. Liege, Mem. (4) 3, No. 1, 1-66 (1939)

Carleman, T. [1] *Sur les systemes linéaires aux dérivées partielles---.* Acad. Sci. Paris, C. R. Hebd., 197, 471-474 (1933)

Cauchy, A. L. [1] *Memoir sur l'emploi du calcul des limites dans l'intégration d'équations aux dérivées partielles. Mémoir sur l'application du calcul des limites a l'intégration d'un systeme d'équations aux dérivées partielles. Mémoir sur les intégrales des systemes d'équations différentielles et aux dérivées partielles, et sur le développement de ces intégrales en séries ordonnées survant les puissances ascendantes d'un parametre que renferment les équations proposées.* C. R. Acad. Sci. (Paris), XV pp. 25, 44, 85, 101, 141 (1842)

Chandrasekhar, S. [1] *A new type of boundary value problem in hyperbolic equations.* Camb. Phil. Soc., Proc. 42, 250-260 (1946)

Chaundy, T. W. [1] *Partial differential equations with constant coefficients.* Lond. Math. Soc., Proc. 43, 280-288 (1933)

[2] *Linear partial differential equations. II.* Quart. J. Math., Oxford Ser., 11, 101-110 (1940)

Chernoff, H. [1] *Complex solution of partial differential equations.* Amer. J. Math. 68, 455-478 (1946)

Churchill, R. V. [1] *Modern operational mathematics in engineering.* McGraw-Hill, New York, 1944

Cimino, M. [1] *Una soluzione in grande del problema di Cauchy per una particolare equazione in tre variabili, ottenuta con un metode di M. Picone.* Accad. Italia, Cl. Sc. Fis. Mat. Nat., Rend. (7) 2, 800-809 (1941)

Cinquini-Cibrario, M. [1] *Intorno ad un sistema di equazioni alle derivate parziali del primo ordine.* R. Ist. Lomb. Sc. Lett., Cl. Sc. Mat. Nat., Rend. (3) 7 (76), 177-184 (1943)

[2] *Sul problema di Goursat per le equazioni del tipo iperbolico non lineari.* Ann. Mat. Pura Appl. (4) 21, 189-229 (1942)

[3] *Sopra alcune questioni relative alle equazioni del tipo iperbolico non-lineari.* Ann. Mat. Pura. Appl. (4) 23, 1-23 (1944)

[4] *Sul problema misto per l'equazione del tipo iperbolico non-lineari.* R. Ist. Lomb. Sc. Lett., Cl. Sci. Mat. Nat., Rend. (3) 7 (76), 247-255 (1943)

[5] *Sopra alcune questioni relative ad equazioni ellittico-paraboliche del secondo tipo misto.* Accad. Sc. Torino, Cl. Sc. Fis. Mat. Nat., Atti 77, 365-383 (1942)

[6] *Teoria della caracteristiche per equazioni non-lineari di ordine n di tipo iperbolico*, Ann. Mat. Pura. Appl. (4) 26, 97-117 (1947)

[7] *Sopra un nuovo problema a limit per un sistema di equazions alle derivate parziali.* R. Ist. Lomb. Sc. Lett., Cl. Sc. Mat. Nat., Rend. (3) 10 (79) 103-111 (1946)

[8] *Sopra il problema di Cauchy per i sistemi di equazioni alle derivate parziali del primo ordine.* Sem. Mat. Univ. Padova, Rend. 17, 75-96 (1948)

Cioranescu, N. [1] *La résolution du probleme de Cauchy pour un systeme d'equations du second ordre par la méthode de Riemann.* Math. Zeit. 32, 481-490 (1930)

Columbo, B. [1] *Sopra un equazione a derivate parziali del quarto ordine.* R. Accad. Naz. Lincei, Cl. Sc. Fis. Mat. Nat., Rend (6) 16, 291-296 (1932)

Courant, R., Friedrichs, K. O., Lewy, H. [1] *Über die partielle Differentialgleichungen der mathematischen Physik.* Math. Ann. 100, 32-74 (1928)

Courant, R., Hilbert, D. [1] *Methoden der Mathematischen Physik.* Julius Springer, Berlin, 1937

Coutrez, R. [1] *Sur les variétés caractéristiques des équations aux dérivées partielles du second ordre.* Acad. Roy. Belg., Cl. Sci., Bull. (5) 28, 266-282 (1942)

[2] *Sur les variétés caractéristiques des équations aux dérivées partielles d'ordre quelconque.* ibid, (5) 29, 15-30 and 71-78 (1943)

Cramlet, C. M., Muggli, E. C., Zuckerman, H. S. [1] *On systems of partial differential equations.* Univ. Wash. Publ. Math. 3, No. 1, 45-54 (1948)

DeDonder, T. H. [1] *Simplification de la méthode d'intégration d'Hadamard.* Acad. Roy. Belg., Cl. Sci. Bull. (5) 763-765 (1943)

von Deuffer, H. [1] *Über die Bernsteinsche Theorie der partiellen Differentialgleichungen.* Univ. Berlin, Schr. Math. Sem. u. Inst. Angew. Math. 2, 237-270 (1935)

Digel, E. [1] *Über die Bedingungen der Existenz der Integrale partieller Differentialgleichungen erster Ordnung.* Math. Zeit. 44, 445-451 (1938)

[2] *Über die Existenz von Integralen der partiellen Differentialgleichung* $fz_x+gz_y=0$ *in der Umgebung eines singulären Punktes.* Math. Zeit. 42, 231-237 (1937)

Doetsch, G. [1] *Les équations aux dérivées partielles du type parabolique, (Conferences internationales sur les équations aux dérivées partielles,* Geneve, 17-20 VI. 1935) Enseign. Math. 35, 43-87 (1936)

[2] *Theorie und Anwendung der Laplace Transformation.* Julius Springer, Berlin, 1937

Einaudi, R. [1] *Sul problema di Cauchy per equazioni differenziali a efficienti singolari.* R. Accad. Naz. Lincei, Cl. Sc. Fis. Mat. Nat., Rend. (6) 22, 492-497 (1935)

Emmons, H. W. [1] *The numerical solution of partial differential equations.* Quart. Appl. Math. 2, 251-257 (1944)

Fantappie, L. [1] *Sulla soluzioni del problema di Cauchy* .... Pont. Acad. Sc. Rome, Ann. ns 3, 403-468 (1939)

Feller, W. [1] *Zur Theorie der Stochastuchen Prozesse (Existenz-und Eindentigkeitssatze.* Math. Ann. 113, 113-160 (1936)

Fox, L. [1] *Solution by relaxation methods of plane potential problems with mixed boundary conditions.* Quart. Appl. Math. 2, 251-257 (1944)

Frankl, F., [1] *On Cauchy's problem for partial differential equations of mixed elliptico - hyperbolic type with initial data on the parabolic line.* Akad. Nauk. S.S.S.R., Izvestiia, Ser. Mat. 8, 195-224 (1944); ibid, 10, 135-166 (1946)

Friedrichs, K. I. [1] *Der Verallgemernerung der Riemannschen Methoden.* Gottingen Nach. 172-177 (1927)

Friedrichs, K. O., Lewy, H.[1] *Das Anfangswertproblem einer nichtlinearen hyperbolischen Differentialgleichung in zwei Veranderlichen.* Math. Ann. 99, 200-221 (1928)

[2] *Über die Eindentigkeit und das Abhangigkeitsgebiet der Losungen beim Anfangswertproblem linearer hyperbolischer Differentialgleichungen.* Math. Ann. 98, 192-204 (1927)

Geheniau, J. [1] *Sur les formules d'Hadamard rélatives aux équations aux dérivées partielles.* Acad. Roy. Belg., Cl. Sci, Bull. (5) 29, 591-607 (1943)

Germay, R. H. J. [1] *Intégration par approximation successive des équations aux dérivées partielles.* Soc. Roy. Sc. Liege, Mem. (3) 12, No. 14 (1924)

[2] *Sur l'intégration par approximations successives des systemes en involution d'équations simultanées, an dérivées partielles du premier ordre, a une fonction inconnue.* Soc. Roy. Sc. Liege, Bull. 7, 525-534 (1938)

[3] *Sur les fonctions de Riemann associées aux systemes d'equations aux dérivées partielles --- du second ordre---.* Soc. Roy. Sc. Liege, Mem. (3) 14, No. 11 (1928)

[4] *Sur le calcul par approximation successive des integrales de certoines équations aux dérivées partielles.* Soc. Roy. Sc. Liege Bull. 1, 36-39 (1932)

Gevrey, M. [1] *Sur les équations aux dérivées partielles du type parabolique.* J. Math. Pures Appl., Paris, (6) 9, 305-471 (1913); ibid, 10, 105-148 (1914)

[2] *Systemes d'équations aux dérivées partielles du type parabolique.* Acad. Sc. Paris, C. R. Hebd. 195, 690-692 (1932)

Giraud, G. [1] *Sur quelques problemes de Dirichlet et de Neumann.* J. Math. Pures Appl., Paris (9) 11, 389-416 (1932)

[2] *Nouvelle méthode pour traiter certains problemes **relatifs** aux équations du type elliptique.* ibid, 18, 111-143 (1939)

Gogoladze, V. [1] *Cauchy's problem for a "Generalized Wave Equation".* Acad. Sc. U.R.S.S., C. R. (Doklady) 1, 169-304 (1934)

Goursat, E. [1] *Lecons sur l'intégration des équations aux dérivées partielles du premier ordre.* Paris 1921. Second edition

[2] *Cours d'analyse matématique.* Paris 1927, Third edition

Green, A. E. [1] *Double Fourier series and boundary value problems.* Camb. Phil. Soc., Proc. 40, 222-228 (1944).

Gross, W. [1] *Bemerkungen zur Existenztheorie bei den partiellen Differentialgleichungen erster Ordnung.* K. Akad. Wiss. Wien, Matn.-nat. Kl., Sitz (8) 123, 2233-2251 (1914)

Hadamard, J. [1] *Lectures on Cauchy's problem.* Yale Univ. Press, New Haven, Conn. 1923

[2] *Le cas hyperbolique.* Enseign. Math. 35, 5-42 (1936)

Hamburger, H. [1] *Über die partiellen linearen homogenen Differentialgleichungen zweiter Ordnung von hyperbolischer Typus.* Math. Ann. 105, 438-493; ibid, 106, 500-536 (1936)

Heins, A. E. [1] *Note on the equation of heat conduction.* Amer. Math. Soc., Bull. 41, 253-258 (1935)

Hopf, E. [1] *Über den funktionalen, insbesondere den analytischen, Charakter der Losungen Elliptischer Differentialgleichungen zweiter Crdnung.* Math. Zeit. 34, 194-233 (1931)

Ignatovskij, V. S. [1] *Zur Laplace - Transformation. III.* Acad. Sci. U.R.S.S., C. R. (Doklady) (N.S.) 4, 107-110 (1936)

Ingersoll, B. M. [1] *An initial value problem for hyperbolic differential equations.* Am. Math Soc., Bull. 74, 1117-1124 (1948)

Janet, M. [1] *Les systemes d'équations aux dérivées partielles.* Gauthier-Villars, Paris, 1927

John, F. [1] *Linear partial differential equations with analytic coefficients.* Nat. Acad. Sc. U.S.A., Proc. 29, 98-104 (1943)

Kamke, E. [1] *Differentialgleichungen reeller Funktionen.* Chelsea Pub. Co., New York, 1947

[2] *Über die partielle Differentialgleichung* $fz_x+gz_y=h$. Math. Zeit. 41, 56-66 (1936)

[3] Same title as [2]. ibid, 42, 287-300 (1936)

[4] *Bemerkungen zur Theorie der partiellen Differentialgleichungen erster Ordnung.* Math. Zeit. 49, 256-284 (1943)

Karimov, D. J. [1] *Sur les solutions périodiques des équations differentielles non-linéaires du type parabolique.* Acad. Sci. U.R.S.S., C. R. (Doklady) (N.S.) 54, 293-295 (1946)

Kellogg, O. D. [1] *Foundations of Potential Theory.* Julius Springer, Berlin, 1929

Kourensky, M. [1] *A method of integrating the general form of a system of partial differential equations of first order in two independent and two dependent variables.* Lond. Math. Soc., Proc. 31, 407-416 (1930)

[2] *L'intégration des équations aux dérivées partielles du séconde ordre avec deux fontions de deux variables indépendantes. I., II., III.* R. Accad. Naz. Lincei, Cl. Sc. Fis. Mat. Nat., Rend. (6) 16, 496-499, 567-571, 612-616 (1932); ibid 17, 53-57 (1933)

Kowalewsky, S. [1] *Zur Theorie der partiellen Differentialgleichungen.* Four. fur die reine u. angew. Math. 80, 1-32 (1875)

Kryzanski, M. and Schauder, J. [1] *Quasilineare Differentialgleichungen zweiter Ordnung von hyperbolischen Typus.* Studia Math. 6, 162-189 (1936)

Lahaye, E. [1] *Sur l'applicotion de la méthode des approximations successives a la résolution des equations aux dérivées partielles linéaires du seconde ordre.* Acad. Roy. Belg., Cl. Sc., Bull (5) 27, 537-551 (1941)

Lednev, N. A. [1] *A new method for the solution of partial differential equations.* Mat. Sbornik 22 (64) 205-226 (1948)

Leray, J. [1] *Les problemes non-linéaires.* Enseign. Math. 34, 139-149 (1935)

Levi, B. [1] *Systems of analytic equations;....* Rosario, (Argentina), Univ. Nac. Lit., Fac. Ci. Mat. Fis.-Quim. Nat. Appl. Ind., Monografias No. 1, 1944.

Levi, E. E. [1] *Sul equazioni del calore.* R. Accad. Naz. Lincei, Cl. Sc. Fis. Mat. Nat., Rend. (5) 16, 450-456 (1907)

Lewis, D. C. Jr. [1] *Infinite systems of ordinary differential equations with applications to certain second order partial differential equations.* Am. Math. Soc., Trans 35, 792-823 (1933)

Lewy, H. [1] *Über das Anfangswertproblem einer hyperbolischen nichtlinearen partiellen Differentialgleichung zweiter Ordnung mit zwei unabhangigen Veranderlichen.* Math. Ann. 98, 179-191 (1927)

[2] *Sur une nouvelle formule dans les équations linéaires elliptiques---.* Acad. Sc. Paris, C. R. Hebd. 197, 112-113 (1934)

Lowan, A. N. [1] *On wave motions for infinite domains.* Phil. Mag. (7), 26, 340-360 (1938)

[2] *On wave motion for sub-infinite domains.* Phil. Mag. (7) 27, 182-192 (1938)

MacDuffee, C. C. [1] *A recursion formula for the polynomial solutions of a partial differential equotion.* Am. Math. Soc., Bull. 42, 244-247 (1936)

Mangeron, D.d. [1] *Sopra un problema al contorno per un equazione differenziale non lineare alle derivate parziali di quarta ordine con le caratteristiche reali da pie.* R. Accad. Nag. Lincei, Cl. Sc. Fis. Mat. Nat., Rend (6) 16, 305-310 (1932)

Martin, L. [1] *Sur les problemes aux limites rélatifs a certains systemes d'équations aux dérivées partielles.* Acad. Roy. Belg., Cl. Sci., Bull. (5) 22, 533-539 (1936)

Mathisson, M. [1] *Eine neue Losungsmethode fur Differentialgleichungen von normalem hyperbolischem Typus.* Math. Ann. 107, 400-419 (1932)

Michlin, S. G. [1] *Problemes aux limites fondamentaux de l'équation des ondes.* Acad. Sci. U.R.S.S., C. R. (Doklady) (N.S.) 29, 281-285 (1940)

Mikeladze, S. E. [1] *On the question of numerical integration of a partial differential equation by means of nets.* Akad. Wiss. U.S.S.R., Georg. Abt., Mitt. 1, 249-254 (1940)

[2] *Sur l'intégration numerique d'equations differentielles aux dérivées partielles.* Acad. Sci. U.R.S.S., Bull. (6) 819-841 (1934); also Acad. Sci. U.R.S.S., C. R. (Doklady) (N.S.) 14, 177-179, 181-182 (1937)

Mitrinovitch, D. S. [1] *Sur l'integration d'une équation linéaire aux dérivées partielles.* Acad. Sci. Paris, C. R. Hebd. 210, 783-785 (1940)

Nagumo, M. [1] *Über das Anfangswertproblem partieller Differentialgleichungen.* Jap. J. Math. 18, 41-47 (1942)

Perron, O. [1] *Ein neuer Beweis des Fundamentalsatzes in der Theorie der partiellen Differentialgleichungen erster Ordnung.* Math. Zeit. 5, 154-160 (1919)

[2] *Über Existenz und Nichtexistenz von Integralen partieller Differentialgleichungsststeme in reellen Gebiete.* Math. Zeit. 27, 549-564 (1928)

Petrowsky, I. G. [1] *New proof of the existence of a solution of Dirichlet's problem by the method of finite differences.* Uspekhi Mat. Nauk 8, 161-170 (1941)

[2] *Zur ersten Randwertaufgabe der Warmeleitungsgleichung.* Comp. Math. 1, 383-419 (1934)

Picard, E. [1] *Lecons sur quelques problemes aux limites de la théorie des équations différentielles.* Gauthier-Villars, Paris, 1930

Picone, M. [1] *Nuove metodi resolutivi ---.* Accad. Sci. Torino, Cl. Sc. Fis. Mat. Nat., Atti 75, 413-426 (1940)

Piskounov, N. S. [1] *Problemes aux limites pour les équations du type elliptico - parabolique.* Mat. Sbornik (N.S.) (7) 49, 385-424 (1940)

[2] *Intégration des equations de la théorie des couches frontieres.* Accad. Sci. U.R.S.S., Bull. Ser. Math. 7, 35-48 (1943)

Raynor, G. E. [1] *Dirichlet's problem.* Ann. Math. 23, 193-197 (1921)

Rellich, F. [1] *Zur Konstruktion der Grundlosung fur eine gemischte Randwertanfgabe einer partiellen Differentialgleichung hoherer Ordnung.* Math. Ann. 112, 490-492 (1936)

Riquier, C. [1] *Les méthodes des fonctions majorantes et les systemes d'équations aux dérivées partielles.* Memorial des Sciences Math. 32, Paris, 1928

Rodabaugh, L. D. [1] *The partial differential equation* $z_x+fz_y=0$. Duke Math. J. 6, 362-374 (1940)

Rosenblatt, A. [1] *Sopra le equazione m-armoniche non lineari due variabili independenti. I.* R. Accad. Naz. Lincei, Cl. Sc. Fis. Mat. Nat., Rend. (6) 19, 212-219, 306-310 (1934)

[2] *Sur l'application de la méthode des approximations successives---.* Acad. Sc. Paris, C. R. Hebd. 196, 460-462 (1934); ibid 197, 1021-1023 (1934); ibid 197, 1278-1280 (1934); ibid 198, 320-322 (1934)

Rothe, E. [1] *Über die Grundlosung bei parabolischen Gleichungen.* Math. Zeit. 33, 488-504 (1931)

[2] *Über einige Analogien zwischen linearen partiellen und linearen gewohnlichen Differentialgleichungen.* Math. Zeit. 27, 76-86 (1928)

Sato, T. [1] *Sur l'équation aux dérivées partielles hyperbolique* $s = f(x,y,z,p,q)$. Memoirs of the Fac. of Sc., Kyusyu Univ., Ser. A, 2, 107-123 (1943)

Schauder, J. [1] *Das Anfangswertproblem einer quasilinearen hyperbolischen Differentialgleichung zweiter Ordnung in beliebiger Anzahl von unabhangigen Veranderlichen.* Fund. Math. 24, 213-246 (1935)

[2] *Numerische Abschatzungen in elliptischen linearen Differentialgleichungen.* Studia Math. 5, 34-42 (1934)

[3] *Sur le probleme de Dirichlet généralisé pour les équations non linéaires du type elliptique.* Acad. Sc. Paris, C. R. Hebd. 195, 201-203 (1932)

[4] *Über den Zusammenhang zwichen der Eindeutigkeit und Losbarkeit partieller Differentialgleichungen zweiter Ordnung vom elliptischen Typus.* Math. Ann. 106, 661-721 (1932)

[5] *Equations du type elliptique, problemes linéaires.* Enseign. Math. 35, 126-139 (1936)

[6] *Zur Theorie stetiger Abbildungen in Funktionalraume* Math. Zeit. 26, 47-65 (1927)

Schroder, K. [1] *Zur Theorie der Randwertaufgahen der Differentialgleichung* $\Delta\Delta U = 0$. Math. Zeit. 48, 553-675 (1943)

Siddiqi, M. R. [1] *Cauchy's problem in non-linear partial differential equations of hyperbolic type.* Camb. Phil. Soc., Proc. 31, 195-202 (1935)

Simonoff, N. [1] *Über die erste Randwertaufgabe der nichtlinearen elliptischen Gleichung.* Univ. Moscow, Bull. Math. 2, No. 1, 18 pp. (1939)

Sjostrand, O. [1] *Sur une équation aux dérivées partielles du type composite.* Ark. Mat. Astr, Fys. 25A, No. 21; 26, No. 1 (1936)

Soboleff, S. [1] *Sur le probleme de Cauchy pour des équations quasi-linéaires hyperboliques.* Acad. Sci. U.R.S.S., C. R. (Doklady) (N.S.) 20, 79-83 (1938)

Southwell, R. V. [1] *Relaxation Methods in Theoretical Physics.* Oxford Univ. Press, Oxford, 1946

[2] *Relaxation methods applied to engineering problems.* Roy. Soc. London, Phil. Trans. ser.A, 239, 419-578 (1943)

Sundaram, M. S. [1] *On non-linear partial differential equations of the hyperbolic type.* Indian. Acad. Sci., Proc. A 9, 495-503 (1939)

[2] *On non-linear partial differential equations of the parabolic type.* Indian Acad. Sci., Proc. A 9, 479-494 (1939)

[3] *Fourier ansatz and non-linear parabolic equations.* Indian. Math. Soc., Jour. (N.S.) 7, 129-142 (1943)

Tamarkin and Feller [1] *Partial Differential Equations.* Brown Univ., Providence, 1941

Tautz, G. [1] *Zur Theorie der elliptischen Differentialgleichungen. I.* Math. Ann. 117, 694-726 (1941)

Taylor, M. [1] *On the uniqueness and existence of solutions---.* Lond. Math. Soc., Proc. ser. 2, 30, 248-263 (1929)

Theodoresco, N. [1] *Les solutions élémentaires d'une classe d'equations aux dérivées partielles linéaires d'ordre supérieur.* Comm. Math. Helv. 10, 164-205 (1937)

Thomas, J. M. [1] *Differential Systems.* Am. Math. Soc., Colloq. Pub. XXI, New York, 1937

Thomas, T. Y. and Titt, E. W. [1] *On the elementary solution of the general linear differential equation of the second order with analytic coefficients.* J. Math. Pures Appl. 18, 217-248 (1939)

[2] *Systems of partial differential equations---.* Ann. Math. (2) 34, 1-80 (1933)

Titt, E. W. [1] *Cauchy's problem for systems of second order partial differential equations.* Ann. Math. (2) 35, 162-184 (1934)

[2] *An initial value problem for all hyperbolic partial differential equations of the second order with three independent variables.* Ann. Math. (2) 40, 862-891 (1939)

Trjitzinsky, W. J. [1] *Analytic theory of singular elliptic partial differential equations.* Ann. Math. (2) 43, 1-55 (1942)

[2] *Analytic theory of parametric linear partial differential equations.* Mat. Sbornik (N.S.) 15 (57) 179-242 (1944)

Vecoua, E. [1] *Romplexe Darstellung der Losungen elliptischer Differentialgleichungen mit Anwendungen auf Randwertprobleme.* Math. Inst. Tblissi, Trav. [Tbliss. Mat. Inst., Trudy] 7, 161-253 (1940)

Waschakidze, D. [1] *Über die numerische Losung der biharmonischen Gleichung.* Math. Inst. Tblissi, Trav. [Tbliss. Mat. Inst., Trudy] 9, 61-73 (1941)

Wazewski, T. [1] *Über die Bedingungen der Existenz der Integrale partieller Differentialgleichungen erster Ordnung.* Math. Zeit. 43, 522-532 (1938)

von Weber, E. [1] *Partielle Differentialgleichungen.* Encykl. Math. Wiss., Bd. II[1], Halfte 1, 294-399, Leipzig (1898-1931)

Winants, M. [1] *Equation hyperbolique du troisieme ordre a coefficients constants et de la catégorie III. Résolution du probleme de Cauchy par la méthode des approximations successives.* Acad. Roy. Belg., Bull. 21, 283-293; 495-503 (1935)

Zwirner, G. [1] *Sugli elementi uniti delle traformazioni---.* R. Accad. Naz. Lincei; Cl. Sc. Fis. Mat. Nat. Rend. (8) 3, 44-49 (1947)

Bieberbach, Ludwig [1] *Theorie der Differentialgleichungen.* Springer, Berlin, 1932; reprinted Dover, N. Y., 1944.

Caratheodory, C. [1] *Variationsrechnung und partiellen differentialgleichungen ersten ordnung.* Teubner, Lepzig, 1935.

Dressel, F. G. [1] *The fundamental solution of the parabolic equation.* Duke Math. Journal, v. 13(1946), 61-70

Engel - Faber [1] *Die Liesche theorie der partiellen differentialgleichungen ersten ordnung.* Teubner, Leipzig, 1932

Holmgren, E. [1] *Sur l'equation de la propagation de le chaleur.* Arkiv for Mat. Astron. och. Fys. 4(1908) No. 14

Saltykow, N. [1] *Etude sur les integrales de S. Lie des equations aux derivees partielles du premier ordre....* Bull. Acad. des Sc. Math. Nat; Acad. Roy. Serbe. Belgrade No. 5(1939) 121-137

Vasilesco, F. [1] *Le Probleme de Dirichlet dan le Cas le plus general.* Enseign. Math. v. 34(1935) 88-106